Village Information System for Development Planning

By the same Author

Village Development Planning

About the Editor

Hridai R. Yadav obtained degrees of Post Graduation, M.Phil., Ph.D. from Jawaharlal Nehru University, New Delhi, worked with CSIR for eight years, having authored 10 books and published 73 research papers. He has been the Member, National Wasteland Development Board, Govt. of India, besides being Founder Member, Secretary, and Director, Institute of Wastelands Reclamation and Rural Development (1990), Institute of Science and Technology for Rural Development (1994) and RIWARD Post Graduate College, Hanumanganj, Sultanpur, U.P. (2004). Dr. Yadav is engaged in the implementation of Women Empowerment, Rural Health and Village Information System Programmes in Amethi. Presently, Dr. Yadav has been appointed as UGC Observer to oversee the selection process from Reader to Professor under CAS Scheme of the UGC, New Delhi.

Village Information System for Development Planning

Innovation and Infrastructure

Edited by

Hridai R. Yadav

Published for

Rajiv Gandhi Institute of Wasteland & Rural Development, Amethi

By

CONCEPT PUBLISHING COMPANY PVT. LTD.
NEW DELHI-110059

ISBN-13: 978-81-8069-942-9

First Published 2013

Published and Printed by

Concept Since 1974

Concept Publishing Company Pvt. Ltd.
A/15-16, Commercial Block, Mohan Garden
New Delhi-110059 (India)
T: 25351460, 25351794, *F*: 091-11-25357109
E: publishing@conceptpub.com, W: www.conceptpub.com
Editorial Office: H-13, Bali Nagar, New Delhi-110 015, India.

Cataloging in Publication Data--*Courtesy:* D.K. Agencies (P) Ltd. <docinfo@dkagencies.com>

International Workshop on "Innovation, Infrastructure and Village Information System for Development Planning" (2011 : Sultanpur, India)
Village information system for development planning : innovation and infrastructure / edited by Hridai R. Yadav.
p. cm.
Includes bibliographical references and index.
ISBN 9878180699429

1. Rural development projects—Geographic information systems—Developing countries—Congresses. 2. Rural development projects—Geographic information systems—India—Congresses. 3. Infrastructure (Economics)—Geographic information systems—Developing countries—Congresses. 4. Infrastructure (Economics) Geographic information systems—India—Congresses. 5. Natural resources Geographic information systems—Developing countries—Congresses. 6. Natural resources—Geographic information systems—India—Congresses. 7. Communication in rural development—Developing countries—Congresses. 8. Communication in rural development—India—Congresses. I. Yadav, Hridai Ram, 1957- II. Title.

DDC 307.1412091724 23

राम नरेश यादव

राजभवन

भोपाल–462 052

9 मई, 2012

संदेश

हर्ष का विषय है कि राजीव गांधी इन्स्टिट्यूट ऑफ वेस्टलेंड एण्ड रूरल डेवलपमेंट द्वारा अन्तर्राष्ट्रीय कार्यशाला में प्रस्तुत शोध, सहमतियों और कार्यवाहियों तथा निष्कर्षों को पुस्तक के रूप में प्रकाशित किया जा रहा है।

राष्ट्रीय और अन्तर्राष्ट्रीय स्तर पर आयोजित कार्यशाला ग्रामीण विकास, महिला सशक्तिकरण और ग्रामीण स्वास्थ्य जैसे मुद्दों पर महत्वपूर्ण मंच सिद्ध हो रहा है। ये कार्यशालाएं राज्य एवं केन्द्र सरकार की नीतियों, नियोजन और क्रियान्वयन में अहम भूमिका निभा रही हैं।

मुझे पूर्ण विश्वास है कि कार्यशाला में विद्वजनों द्वारा पारित प्रस्ताव व संस्तुतियों का प्रकाशन ग्रामीण विकास के क्षेत्र में अत्यधिक सहायक और परिणाममूलक सिद्ध होगा तथा वंचित, पीड़ित, शोषित और असहाय लोगों को देश की मुख्यधारा से जुड़ने का अवसर भी प्राप्त हो सकेगा।

शुभकामनाएं।

(राम नरेश यादव)

राज्यपाल, मध्य प्रदेश

दूरभाष : 0755.4080170, 4080180, फैक्स : 0755.4080172

Preface

Integrated economic development has been realized and commendable advances have been made in the diverse field of rural development programmes, would in reality be incomplete, if such rural development programmes fail to translate into better and dignified life for the majority of people who live in remote rural areas, because the **benefit of development must reach to the down-trodden deprived rural resourceless poors equally, effectively and efficiently** but increasingly there is a talk on:

- Rural Area
- Urban Area
- Rurban Area

It is argued that faster development is required to **bridge the gap** between **rural-urban and rurban area.** An appropriate and faster development of rural areas are required to fulfil the increasing demand of ever growing population, so that the disparities between rural and urban can be bridged to ensure integrated development.

It has been felt that the development should be participatory and it should be planned according to the local needs of the deprived rural multitudes. Development should be planned through tapping the village infrastructure—natural and human resources, local talent in order to ensure implementation of development programmes self-sustainable.

Various rural development programmes are being implemented aiming at sustainable, integrated development of rural areas with the focus on enhancing the livelihood of weaker and vulnerable sections of the society. In conformity with the commitment to accord foremost priority to village development, a number of steps have been initiated to make the implementation of ongoing rural development programme more effective and accelerated.

The formulation and implementation of rural development programme in the village is an **intensive, innovative, informative** and **infrastructure** (Natural and Human Resources) oriented activity. The **natural and human resources infrastructure, information system at village level** are not properly arranged but such available information needs to be **integrated and innovated for village development planning**. The conventional methods of collection of information, collation, storage are not appropriate for quick updation, retrieval and holistic analysis and several problems are existing in the conventional system.

In view of the requirements and development in the areas of Information and Communication Technology (ICT) and Geo-Spatial Data Management and Analysis Techniques like Remote Sensing and Geographic Information System (GIS). Certain initiatives have been taken at the National level to create awareness in the usefulness of these technologies between planners and decision makers.

Natural Resources Data Management System (NRDMS) programme of **Department of Science and Technology**, Government of India has been one such pioneering effort, while others are being:

- National (Natural) Resources Information System (NRIS) of the Department of Space.
- Integrated Mission for Sustainable Development (IMSD) of the Department of Space.
- Geographic Information System (GIS/NIC) and District Information System (DIS/NIC) of the National Informatics Centre.
- National Spatial Data Infrastructure (NSDI) of Department of Science and Technology, Government of India.

In view of the above requirement, Department of Science and Technology, Government of India during 1982 started a programme on Natural Resources Data Management System (NRDMS), which is a multi-disciplinary and multi-institutional research and development programme with a vision to provide appropriate scientific-technological inputs for operationalizing the concept of decentralized planning.

2. Panchayati Raj System

Participatory planning and development should be initiated and sustained through the participation of rural people at grassroots/village level.

It has been mentioned in the Constitution of India that *Government should evolve methods to unite the village Panchayat and provide such powers and rights which can be helpful to work as an unit of Self Government.*

The Father of the Nation Mahatma Gandhi Ji stated, *"In the Panchayati Raj System, each village should be self-sustained and independent, the implementation of the welfare programme of rural masses, can be implemented in a better manner in the village of Vast Rural India."*

In view of the above, as per policy of the Government of India, the State Governments established Self Government to strengthen the Panchayati Raj Institutions in Rural Areas. Historical and revolutionary decison has been taken to allot the work and to allocate financial resources to implement the planned programmes and also allotcated financial, administrative rights to provide the **Self-Governance** of Village Panchayat, Block Panchayat and District Panchayat.

The real meaning of democracy is the importance to the people and Self Government is the main sole of rural development, Management, Financial and Administrative Control of Village Panchayat and continued participation, co-operation of the local people to establish Village Panchyat for **Self Government For the People, By the People, with the People,** to ensure participatory planning, implementation of rural development programmes for self-sustenance.

The optimum utilization of available natural, socio-economic, institutional, infrastructural resources to meet the increasing demand of growing cattle and human population for Fuel wood, Fodder, Fibre, Fruits, Fisheries, Food grains (F^6) and Vegetable should be ensured through **Self Government-Panchayati Raj-Lok Raj, Gram-Raj, Ram-Raj** in which scientific-technical-infrastructure facilities are procured in an innovative manners to generate employment and

Allocated Work for Panchayati Raj Institutions

Village Panchayat

1. Education
2. Drinking Water
3. Irrigation
4. Youth Welfare
5. Medical and Health
6. Women and Child Development
7. Food Supply Shop.
8. Animal and Agriculture
9. Panchayati Raj and Rural Development
10. All types of pension and scholarship

Block Panchayat

1. Rural Development
2. Primary Health Centre
3. Animal Hospital
4. Seed Centre
5. Marketing Store
6. Maintenance of Assets
7. Additional Villages

District Panchayat

1. Medical and Health
2. Development Programme
3. Education.
4. Tube wells Government
5. Youth Welfare.
6. Women and Child Development
7. Animal Department
8. Food Supply Shops
9. Agri.-Horticulture Department
10. Panchayati Raj
11. Jal Nigam
12. Food Supply
13. Forest Department
14. Public Works Department
15. Irrigation Department
16. Industries Department
17. Fisheries Department
18. Other Departments

increase the income of deprived rural multitudes surviving in the remote rural villages, to improve their socio-economic conditions for integrated sustainable development.

In view of the above, an attempt has been made to point out certain innovative, infrastructural resources, i.e. natural and human resources (socio-economic institutional) and the implementation of various rural development programmes being implemented under the various schemes of the Government/Semi-Government/Non-Government organizations. It has also been attempted to explain the need based village development planning indicators which may be helpful in the preparation of the research papers for presentation in the International Workshop.

The Department of Social Work, **RIWARD Post Graduate College,** Hanumanganj, Sultanpur under **Institute of Wastelands Reclamation and Rural Development**, Sultanpur in collaboration with **NRDMS, Department of Science and Technology, Government of India,** New Delhi organized an International Workshop on **10-11 November, 2011** to provide open platform for

Indicators: Innovation, Infrastructure and Village Information System for Development Planning

Theme	*Sub-Theme*
Natural	Natural Resources Physiography, Drainage System, Rainfall, Soil, Land use, Forest and Vegetation.
Social	Introduction, Population—Male and Female, Caste composition, Workers (Primary, Secondary, Tertiary), Educational Profile.
Economic	Cultivation, Land ownership, Agriculture, Fertilizer, Irrigation, Agricultural Production, Animal Husbandry, Primary, Secondary and Tertiary Sectors of Economy.
Infrastructure	Historical infrastructure, Irrigation, Administrative, Road, Communication, Drinking Water Energy, Medical/Health, Agro Industry, Small Scale Industry, Animal Hospital, Banking, Educational Infrastructure.
Implementation of Development Programme	MNREGA Energy, Small Scale Industries, Drinking Water, Sanitation, Rural Housing, Welfare Programme, Road, Women and Child Development Programme, Agriculture, Horticulture, Dairy Development, Medical Health, Education.
Village Development Planning	Problem of the village, Planning for Agriculture, Education, Rural Development, Rural Health, Irrigation and Communication, Solution of the Problem, Action Work Proposed planned action work. Achievements of the Action work, Constraints.
Programme Implementation	Women Empowerment (SHG) National Rural Health Mission Selection (NRHM) of down trodden, deprived, beneficiaries for SHGs and rural health to provide Government facilities and support for employment and income generation to SHGs and NRHM facilities and other programmes, viz. education, rural development, employment, road, drinking water, communication, transportation social forestry, agriculture, horticulture development and other programmes as per need of the deprived rural poor to strengthen the Panchayati Raj System.

discussion and to examine the experiences of the country at village level, i.e. grassroots level on this issue. The challenges/hurdles faced and generated ideas that would help in re-casting the innovative-informative-infrastructural village development planning for sustainable development.

3. Objectives of the Workshop

- To critically examine the substantive scope for information-innovation-infrastructure at grassroots level, i.e. village level.
- To analyze the natural and human resources (Socio-Economic-Infrastructure) and Rural Development Programme Implementation at Village Level.
- To review the Village Information System.
- To critically examine the need based village development planning.
- To review the Panchayati Raj System and the powers, rights, financial, administrative, managerial support system for self-government to ensure sustainable development.
- To identify the challenges, problems and constraints for information, innovation, infrastructure and village development planning.
- To suggest suitable remedies for innovative/information system to collect the available infrastsructure/resources for Village Information System for Development Planning.

International level academicians from social sciences, agriculture, environmental science and concerned government departments including implementing agencies engaged in the implementation of Rural Development of village level, policy makers, planners, administrators at grassroots level and representatives or civil policy, NGOs, Media, social development activists participated in the workshop.

The International workshop was inaugurated by H.E. Sri Ram Naresh Yadav, Hon'ble Governor, Madhya Pradesh on 10.11.2011 at 10.20 AM at RIWARD Post Graduate College, Hanumanganj, Sultanpur (U.P.), India. Sri Govind Raju, N.S. (IAS), District Magistrate, Sultanpur, Prof. R.B.S. Verma and Prof. A.N. Singh, Department of Social Work, Lucknow University, Lucknow, were present in the inaugural session. Recommendations of the International Workshop has been prepared for its onward submission to the Department of Science and Technology, Government of India.

7th June, 2012 **Hridai R. Yadav**

Acknowledgements

The organization of an International Workshop within the limited resources in the remote rural area at RIWARD Post Graduate College, Hanumanganj, Sultanpur cannot be made successful without the continued support of Sri Vilas Rao Deshmukh, Hon'ble Minister of Science and Technology, Government of India, Maj. Gen. (Dr.) R. Shiva Kumar, Adviser and Head, NRDMS and Dr. Bhoop Singh, Director, NRDMS, Department of Science and Technology, Government of India, New Delhi.

We are grateful to Shri G.K. Das, Private Secretary to the President of India for sending the President's wishes "the workshop all success" and Sri Salman Khursheed Hon'ble Minister of Minorities Affairs, Government of India, New Delhi.

It is our pride privilege to acknowledge of gratitudes and thanks to HE Sri Ram Naresh Yadav Ji, Hon'ble Governor, Madhya Pradesh for his kindness to inaugurate this International Workshop on 10-11 November, 2011 forenoon as Chief Guest.

We are grateful to Sri Govind Raju NS (IAS), District Magistrate, Sultanpur and Sri Govind Aggarwal (IPS), Superintendent of Police, Sultanpur for their necessary support for successful organisation of the workshop. Prof. R.B.S. Verma, Prof. A.N. Singh from Lucknow University and Shailendra Kumar Yadav, KGMU, Lucknow deserve heartiest greetings for their support.

Sri Shasi Kant Srivastava, Lucknow University and Sri Kamlesh Chandra Verma deserve heartiest thanks for websites and Email support etc.

We are grateful to the Raj Kumar Mishra, D.K. Tiwari, Santosh Gond and others, who directly or indirectly sùpported for the successful organization of the workshop and we are thankful to Sri Sanjary Aggarwal for printing of workshop material.

Indeed we are grateful to Sri Ashok Mittal, CMD, Concept Publishing Company Pvt. Ltd., New Delhi-110059, Deliberations of the International Workshop for wider publicity at International and International.

Lastly, I would like to thank my beloved wife Smt. Reeta and son Sahit Hridai for their sustained support and inspiration.

Hridai R. Yadav

Contents

सदेश, राम नरेश यादव, राज्यपाल, मध्य प्रदेश के द्वारा	v
Preface	vii
Acknowledgements	xiii
Contributors	xix

SECTION I
INNOVATION OF NATURAL INFRASTRUCTURE

Introduction	1
1. Innovation, Information System for Natural Infrastructure *Hridai R. Yadav*	3
2. Indigenous Knowledge for Resource Management: Perspectives of Jamuna River Chars in Bangladesh *Shaikh Shahriar Mohammad* and *Sarawat Rashid*	13
3. The Current Perception of the Forest Juju in Ecological Conservation in an African Context *Humphrey NGALA NDI*	40
4. Integrating Traditional Knowledge in Climate Change Adaptation *V. Dhulasi Birundha*	53
5. Reclaiming Wastelands for Agro-Afforestation in Amethi *Hridai R. Yadav*	74

6. Resource Utilization of Sundarban Mangrove Forest People: An Ecological Perspective 93
Mohammad Majumdar, S.M. Monirul Hassan and *Rashed-Un-Nabi*

SECTION II
INNOVATION OF SOCIO-ECONOMIC INSTITUTIONAL INFRASTRUCTURE

Introduction 111

7. Innovation, Information System for Social Infrastructure in Village Planning 113
Hridai R. Yadav

8. Impacts of Petroleum Industry on Fishing Communities: A Case Study of the Todos os Santos Bay, Bahia, Brazil 121
de Barros and *Shashi Kant Srivastava*

9. Innovation, Information System for Economic Infrastructure in Village Planning 136
Hridai R. Yadav

10. Sustainable Yogic Farming: A Social View 144
U. Yadav and *Shailendra K. Singh*

11. Innovation, Information System for Institutional Infrastructure for Village Development Planning 154
Hridai R. Yadav

12. 'E-Krishi Kiran'—Village Information System for Scientific Farming 162
A.M. Shekh and *Arun Patel*

SECTION III
RURAL DEVELOPMENT PROGRAMMES

Introduction 173

13. Nation and Rural Development in India: Images of Gandhi and Nehru 175
Rajesh Kumar Nayak

14. Innovation, Information System for Rural Development Programmes 182
Hridai R. Yadav

15. Traditional Leadership in Ghana 190
Jones Lewis Arthur and *Michael Kofi Nsiah*

16. The Role of M.S.W. Students in Empowering Deprived Rural Women for Sustainable Development 210
Hridai R. Yadav

17. Development of India through Integrated Rural Development 231
Vivek Kumar Singh

18. Need Based Village Development Planning 244
Hridai R. Yadav

19. Communication Strategy for Sustainable Development: The Indian Experiment 252
Mahendra Kumar Padhy

20. Role of Community Radio in Rural Development-Based Communication 260
Ravi Suryavanshi and *Neha Nigam*

SECTION IV
RURAL HEALTH MISSION

Introduction 265

21. Planning for Rural Health Mission 267
Hridai R. Yadav

22. Vehicular Pollution: Health Effect and Its Control 275
Manish Yadav and *Uttara Yadav*

23. Role of National Rural Health Mission (NRHM) in Providing Effective Health Care Services to Rural Population in India 287
Rajneesh Chaudhary

24. Role of Availability and Beneficiaries of Inclusive Village Development Programme in Reproductive Health Services in Uttar Pradesh, India 297
Diwakar Yadav and *Shashi Kant Srivastava*

Recommendations 317

Index 319

Contributors

A.M. Shekh, Vice-Chancellor, Anand Agricultural University (AAU), Anand (Gujarat), India.

Arun Patel, Extension Education Institute, AAU, Anand (Gujarat), India.

Diwakar Yadav, Research Scholar, International Institute for Population Sciences (IIPS), Mumabi-400088, India.

H.R. Yadav, Director/Principal—RIWARD P.G. College, Hanumanganj, Sultanpur, U.P. (India).

Henrique de Barros, UFRPE/Fundaj, Recife-Brazil.

Humphrey NGALA NDI, University of Buea.

Jones Lewis Arthur, Lecturer, Research Fellow, Sunyani Polytechnic, Ghana.

Mahendra Kumar Padhy, Deptt. of Mass Communication and Journalism, Babasaheb Bhimrao Ambedkar University (A Central University), Lucknow, U.P.

Manish Yadav, M.Tech. Environmental Engg., MNIT, Jaipur.

Michael Kofi Nsiah, Ph.D. Fellow, University of Victoria, B.C., Canada.

Mohammad Majumdar, University of Beremen, Beremen, Germany.

Neha Nigam, Research Scholar, Department of Mass Communication and Journalism, Babasaheb Bhimrao Ambedkar University (A Central University), Lucknow.

Rajesh Kumar Nayak, Asstt. Professor, Department of History, J.P. University, Chapra (Bihar).

Rajneesh Chaudhary, Research Scholar, Department of Social Work, University of Lucknow.

Rashed-Un-Nabi, Professor, Department of Marine Science, University of Chittagong, Bangladesh.

Ravi Suryavanshi, Research Scholar, Deptt. of Mass Communication and Journalism, Babasaheb Bhimrao Ambedkar University (A Central University), Lucknow.

S.M. Monirul Hassan, Associate Professor, Department of Sociology, University of Chittagong, Bangladesh.

Sarawat Rashid, Research Associate, Research and Evaluation Division, BRAC Centre, 25 Mohakhali, Dhaka-1212.

Shaikh Shahriar Mohammad, Assistant Director, Rural Develoment Academy, Bogra, Bangladesh.

Shailendra K. Singh, Research Scholar (Philosophy), University of Lucknow, Lucknow.

Shashi Kant Srivastava, Department of Social Work, University of Lucknow.

U. Yadav, Associate Professor (Sociology), Mahila P.G. College, Lucknow.

V. Dhulasi Birundha, Senior Professor, Former Head, Deptt. of Environmental Economics, School of Economics, Madurai Kamaraj University, Madurai-625021.

Vivek Kumar Singh, 'Reader', Department of Social Work and Sociology, Dr. Shakuntla Misra Rehabilitation University, Mohan Road, Lucknow.

Section I
Innovation of Natural Infrastructure

Introduction

Innovation of available natural resources at village level and developing information system for micro-level planning plays an important role for integrated development of village. The natural infrastructure available at village level has spatial contribution in the overall economic development of any village at grass- roots level. H.R. Yadav has made an effort to explain the available natural infrastructure, viz. underground water, forest cover, mines, soil, surface water, land use system, quality of water, land capability and other indicators related to natural resources in the village Sarai Khema of Amethi Block, which plays an important role in the formulation of need based village development planning.

Shekh Shahriar Mohammad and Sarawat Rashid has attempted to explain the Indigenous knowledge for resource management. Perspective of Jamuna River has in Bangladesh. The main objectives of this study has been to find out the impact of seasonality on livelihood and to investigate the resource management for coping with seasonal variation of the study people when they faced vulnerabilities and scarcity.

Humphery NGALANDI has studied the current perception of the forest juju in ecological conservation in an African context.

V. Dhulasi Birundha has made an effort to study the integrated traditional knowledge in the climatic change adaptation. Local knowledge system have been found to contribute to sustainability in diverse fields such as bio-diversity, conservation and maintenance of ecosystem, genetic resources conservation and management of other natural resources. Local knowledge system has also been found useful for ecosystem restoration.

H.R. Yadav has explained the declaiming the wasteland for agro-afforestation in Amethi. In this study an attempt has been made to study the reclamation of available wastelands at village level, its distribution, factors responsible for growth of wastelands, practical reclamation and growth of wastelands, practical reclamation and demonstration of wastelands with the people's participation for fuel wood, fodder, fibre, fruits, foodgrains, fisheries and vegetables to meet the increasing demand of growing cattle and human population.

Mohammad Mazumdar, S.M. Monirul Hassan and Rashed-un-Nabi has studied Resource utilization of Sunderban Mangrove Forest People; an ecological perspective, because the exploitation of resources is always sensitive to the existing environment. People and environment do not merely exist side by side for their own respective survival. It has been argued in this study that despite the presence of maximizing resource utilization motivation, people are not indifferent to the destruction of their immediate environment, knowing the critical balance between man and nature, people of Sunderban area are involved in creating balance in it.

1

Innovation, Information System for Natural Infrastructure

Hridai R. Yadav

1. Introduction

Innovation of available natural resources at village level and developing an information system for micro-level planning plays an important role for integrated development of village. The natural infrastructure at village level has spatial contribution in the overall economic development of any village at grassroots level. The underground water, forest covers, mines, soils, surface water, land use system, quality of water, land capability are the major important component of the natural resources infrastructure.

The District Amethi, and Block Amethi is endowed with the natural resources. It has been observed that in some parts of the Amethi natural resources have been a boom for the development programmes, viz. fertile agricultural land while water logging and usar land has led for an increase in the barren land and wasteland causing non-fertility of the land which has been the main reason for the backwardness of Amethi.

In view of the positive and negative aspects of the available natural resources/infrastructure of Amethi Block, it is an important task to study and analyze the available natural infrastructure of village Sarai Khema which plays an important role in the integrated development and planning of Amethi.

2. Objectives

The main objectives of the study are as follows:

- To innovate the available natural resources of the village Sarai Khema of Amethi Block.
- To analyzc the land use system of village Sarai Khema.
- To explain the available waste lands of the village.
- To analyze the surface and underground water supply of the village Sarai Khema.
- To study the soil system of the village and land capability.
- To analyze the forest and vegetation covers of the village Sarai Khema.

3. Innovation of Natural Infrastructure in the Village Sarai Khema

Available natural resources has an important contribution in the integrated economic development of any area. The underground water/ surface water, forest cover, soil, drainage system, land use are the main important components of the natural resources. Village Sarai Khema of the Amethi Block is endowed with natural resources. In some parts of the village the natural resources has been boon for the economic development having very fertile agricultural land while, water-logging and land salinity have caused an increase in the barren and wastelands leading towards non-fertility of the available land resources, which has been the main reason for the backwardness of the villages of Amethi Block.

The innovation of available natural resources has been ensured for the agro-climatic conditions, physiography of land use, soil, drainage system, vegetation and forest cover at village level in Amethi Block. The details of the natural infrastructure of the village Sarai Khema of Amethi Block has been analyzed.

Historically the past of the Amethi had been extremely famous, honoured, appreicated archaeologically, culturally, geographically. Amethi has very important place in the history. In spite of the several continued changes identity of the Amethi has been permanent.

4. Geology

There are no noteworthy feature in the physiognomy of the area. Geologically Amethi is not important. Geology of the area is characterised by a unrelieved monotony of alluvial deposits, which completely shroud the old land surface to a depth of several metres. Kankar land is found in the area. The soil of the area is formed of rich alluvial soil which has great agricultural importance but kankar or calcareous lime stone are common.

Geology of the Amethi does not reveal anything striking except the ordinary Gangestic alluvium. The kankar pan is found at depth up to about one metre from the surface. Reh and Multani Mitti known as clay is also found in some parts of the area, older alluvial, saline and saline/alkaline soils are found in the area.

5. Physiography

Amethi has no physiographic importance. There is negligible availability of rocks and minerals. The physiography of the area is gradual with a scarcely perceptible slope from north-west to south-west. The surface of the area is generally level. The average height of the area is 110 metres. Heterogeneity of the physical landscape is caused actually by local eminences. There are perceptible notches of banjar tracts. The area has been filled by the aluvium brought down by the Himalayan Rivers. Actually the depth of the plain is not known. Boring down mainly for artesian wells has penetrated only up to 1606 metres is the recent alluvium strata. Geologists differ in their estimates of the thickness of the alluvial deposits on the basis of observations.

6. Drainage System

Amethi is a part of an integrated drainage system of the river Chamraura, although different Nalas join the stream. In general drainage pattern of the area is dendritic and the general characteristic features throughout the plains are that the rivers meet at the acute angles and several nalas from parallel lines of the main stream. There are several small streams. The Chamraura river flows from north-west to south-east. There are tanks, *jhils*, and lakes in Amethi. There are number of shallow *jhils* and swamps but none of them are of any considerable size or importance.

7. Climate

Amethi experiences a rhythm of seasons with reversal of prevailing winds. The reversal is regularly left twice during the year from

November to May winds blow from the west and are of the continental character whereas from June to October they are of the oceanic origin and blow from the east. The complete scale reversal of direction of winds on a gigantic scale between summer and winter is responsible for the monsoon season.

Amethi experiences typical tropical climate with distinctive summers, rainy and winter seasons. The mean daily temperature rises to 40.6°C and mean daily minimum temperature dropped to 8.6°C on an average. The avverage annual rainfall is 1005 mm. Occurrence of dust-storm and thunder storms in the hot season is the special phenomenon of the local climate in the area. The relative humidity rises to 85 per cent during the south-west monsoon and decreases to 30 per cent in the summer season. The climate of the area is normal and suitable for agricultural activities.

8. Soils

Soils of Amethi are transported from sedimentary rocks of the Himalay as and the product of decomposition of these rocks have been deposited over the tract through water action. A brief account of the soils is, however, prepared by the revenue authorities to assess the land revenue. The main factors governing the system of classification are texture colour, availability of water and the level of the land.

The soil characteristics including physical and chemical characteristics of the area is general. Mechanical component of the soil has been recorded 57.6 per cent sand, 17.3 per cent silt and 24.1 per cent clay is found in the soil while textural class of the soil has been found from loam to silt clay loam. The *p*H value of the soil has been recorded as 8.6. The soils of the area is found medium to fine texture.

Representative soil profiles of the village is as follows:

00-30 cm: Silt clay loam; light brown low permeability, roots, sparse, iron and lime granules not apparent in the soil.

30-50 cm: Silt loam, yellowish brown, sticky when wet and very hard on drying, low permeability.

50-90 cm: Silt loam, yellowish brown, angular blocky, sticky when wet and very hand on drying, low permeability, fine granules of iron and lime moderate effervescence.

90-130 cm: Silty clay loam, grey brown, angular blocky, sticky when wet and very hard on drying; low permeability no roots, granules of iron and lime.

130-150 cm: Silty clay, grey, angular blocky, sticky when wet and very hard on drying, low permeability, no rots granules of iron and lime are found.

The soil is deep in nature and inadequately drained having low permeability and saline/alkaline soil fertility of area has been evaluated for micro nutrients (Nitrogen, Phosphorus and Potash) and micro-nutrients (zinc, copper, iron, manganese) to assess the fertility status of the soil.

The macronutrients of the soil, viz. Nitrogen—1.8, Phosphorous—1.6 and Potash—3.1 has been recorded while micro-nutrients, viz. zinc ppm—0.50, copper ppm—1.36, iron ppm—12.65 and manganese—43.05 have been found in the soils of the area.

The land of the area have limitation to crop productivity due to marked drainage impedance and marked depth limitations to crop rooting in sub-soil kankar layers and soil sodicity singly and in combinations.

Saline and alkaline soils are found in the area, which has excessive proportions of soluble salts and for exchangeable sodium due to which crop production is seriously hampered are locally known as Reh or Usar. Such soils have high *p*H value. There is deficiency of calcium, nitrogen, micro-nutrients like zinc, extremely low water intake rate and hydraulic conductivity, resulting in poor moisture transmission characteristics. The presence of calcium carbonate nodules called Kankar often occurring at almost one metre depth and poor physical condition leads to hindrance on root penetration.

9. Land Use

The land use system for the different categories varies from one part to another due to varying agro-climatic conditions, soil type and applicability as per needs of the local people. The total reporting area of village Sarai Khema is 557 hectares.

Land Use: Village Sarai Khema, Block Amethi, District Amethi

Sl. No.	*Land Use*	*Area in Hectares*
1.	Total Reporting Area	557
2.	Settlement Road etc.	45
3.	Kabristan land	02
4.	Other uncultivated land excluding fallow land	18
5.	Water logged land	29
6.	Area not available for cultivation	94
7.	Horticultural Land	07
8.	Permanent Pasture Land	01
9.	Banjar Land	06
10.	Old Fallow Land	14
11.	Fallow Land	36
12.	Current Fallow Land	43
13.	Net Area Sown	207
14.	Total Irrigated Area	300

Area under settlement, road etc. is 45 hectares and 2 hectares land is earmarked for Kabristan in village Sarai Khema. Other uncultivated lands excluding fallow land is 18 hectares in the village. Area under waterlogged land in the village Sarai Khema is 29 hectares while area not available for cultivation is 94 hectares which has to be brought under cultivation as per needs of the local village people. Horticulture land in village Sarai Khema is 7 hectares and permanent pasture land is 1 hectare. The area under banjar land is 6 hectares while 14 hectares land is left as old fallow land in village Sarai Khema. There are 36 hectares land as fallow land while 43 hectares land is left as current fallow for restoration of the fertility status of the soil. The other cultivable land is 107 hectares in village Sarai Khema of Amethi Block.

The net area sown in village Sarai Khema is 207 hectares lands while total irrigated area is 300 hectares in village Sarai Khema of the Amethi Block, Canal irrigated area is 11 hectares, government tube well irrigated area is 36 hectares and private tube well irrigated area is 248 hectares in village Sarai Khema. There are 56 hectares unirrigated land while double cropped multi-irrigated area is 183 hectares.

It has been observed that 84.5 per cent land holders are owning less than one hectare land while 9.8 per cent land holders are owning 1-2 hectares land. Land size between 2-3 hectares land is owned by 3.0 per cent land holders while 1.9 per cent land holders own 3-5 hectares land size holdings and 8 per cent of land holders own land size above 5 hectares in village Sarai Khema of Amethi Block. Thus it is evident that low size of land holding below one hectare land is owned by 84.5 per cent land holders in village Sarai Khema of the Amethi Block.

10. Wastelands

The wastelands are such lands which are lying uninhibited and uncultivated which is no longer surving any purpose or the land which is left out of cultivation to maintain the fertility status of the soil which termed as fallow land and after a year if not cultivated, the left out uncultivated land is termed as old fallow land and after successive years if the same land is not being brought under cultivation such uncultivated land is termed as banjar land.

The wastelands are such lands which has been previously used but it has been abandoned and no further use has been found for it.

There are 11.36 per cent wastelands in Amethi Block. The water logged land is 0.47 per cent, usar land 1.89 per cent, Banjar land 1.13 per cent, old fallow land 2.56 per cent, fallow land 3.85 per cent and other types of wastelands are 0.35 per cent in Amethi Block of Amethi District.

There are 13.35 per cent waste lands in Sarai Khema village of Amethi Block. There are 0.15 per cent waterlogged land, usar land 3.21 per cent, Banjar Land 1.22 per cent, old fallow land 1.85 per cent, fallow land 7.22 per cent and other types of wastelands 1.54 per cent in Sarai Khema Village of Amethi Block.

11. Natural Vegetation

The village Sarai Khema and its revenue villages has an almost inhabited human occupance for over three million and century of plough, pastoral, culture has reduced the natural vegetation, except in pockets. Suitable agro-climatic condition, fertile soil, flora presents no peculiarities, patches of Dhak are found in the village. There are well wooded and containing magnificent groves of mango, jamun

(*Eugenia Jambolana*) and Mahua (*Bassia Latifolia*) trees in village Sarai Khema of the Amethi Block.

Forest is natural resource and it plays an important role in maintaining the ecological balances of the area, in providing raw materials to cottage and small scale industry, firewood and fuel to the down-trodden section of the society, grazing land, fodder to the animals. Horticultural land in Sarai Khema village is 7 hectares having mango, guvava and other trees.

12. Ground Water

The main source of irrigation and drinking water is underground water. which accounts more than 60 per cent of irrigated areas of the village. Due emphasis has been given to the development of groundwater resource. It has been recorded that 36 hectares land is being irrigated by the Government tube wells owned by the individual farmers, while only 11 hectares land is being irrigated by the canal irrigation system in village Sarai Khema of Amethi Block. There are large number of tube wells, open wells dug-cum-borewells have been installed in the village Sarai Khema. The movement and occurrence of ground water depends on the hydro geological characteristics of the sub-surface formation. The aquifer zone extends to depths greater than 100 m below the ground surface. The transmissivity values varies in the range of 1000 to 2000 m^2/day. Annual average rainfall in the area is about 1000 mm which is the principal source of ground water recharge. Canal irrigated water and wells are the another major source of underground water recharge.

Rainfall and seepage from canal irrigation system and irrigation return flows constitute the principal important sources of underground water recharge. It has been estimated that recharge from rainfall is 20 per cent and recharge due to seepage from canals is 1.8 to 2.5 cumecs/million square metres of wetted area and return seepage from irrigation fields under canal irrigation is 35 per cent of water delivered at outlet. The ground water recharge through return flow from irrigation field underground water irrigation is 30 per cent of water delivered at outlet. The ground water recourse of the village is enough to fulfil the irrigation and other requirements of the local village people.

The waterlogged land of the village Sarai Khema is 29 hectares which is another source of underground recharge.

Thus the ground water has been the major source of irrigation in the village Sarai Khema followed by nominal canal irrigation. There is ample scope for further development and utilization of underground water resources.

13. Forest

Forest being the natural infrastructure/resources plays an important role in maintaining the ecological imbalances of the area and in providing fuel wood, and raw material to the cottage industry and grazing land and fodder to the animals. There is no reserve forest in the village Sarai Khema. Afforestation and social forestry plantations have been implemented along both sides of road, chakroads, canals and village Panchayats common land. Afforestation and agro-forestry is an important programme of any action plan for the integrated development of village Sarai Khema of the Amethi Block. Afforestation is helpful in sustaining and maintaining the ecological balance and it is also helpful in fulfilling the several local people's increasing demand.

The multiple use of trees, leguminous tree species are preferred under social forestry programmes, by the villagers to fulfil the increasing demand of growing cattle and human population.

Important tree species for planting under social-forestry and agro-forestry programmes are—Mahua (*Madhuca longifolia*), Jamun (*Syzygium cumini*), Mango (*Mangifera indica*), Kanji (*Pangamia Pinnata*), Eucalypus, casuarina, shisham (*Dalbergia sissoo*), Arjun (*Terminalia arjun*), common species planted by the local village people to meet the increasing demand of growing cattle and human population for fuel wood, fodder, fibre to fruits etc.

In general village Sarai Khema lacks forest and vegetation because most of the area is brought under cultivation. The orchards and groves comprises only 7 hectare of land. The forest and vegetable cover which is low in terms of area have an important share in the per capita income of the area. Overall the village Sarai Khema is endowed with natural resources which has positive impact on the economic development of the area and local people.

14. Land Resources

Natural land resources of Village Sarai Khema indicate that there are 207 hectares cultiable land in village Sarai Khema and 94 hectares land is not available for cultivation, usar land 8 hectares, 29 hectares of land is waterlogged land, under uncultivated land is 18 hectares, area under settlement is 45 hectares.

The banjar land area is 6 hectares, old fallow land is 14 hectares, fallow land 36 hectares and current fallow land is 43 hectares in village Sarai Khema.

Thus it is evident that if the cultivable waste land, other uncultivated land, area not available for agriculture, Banjar land, waterlogged land, old fallow land, fallow land and current fallow land which is 234 hectares land is developed to fulfil the increasing demand of growing population may be an asset for the integrated development of the village Sarai Khema.

15. Conclusion

It is known fact that the village Sarai Khema is endowed with the natural resources and natural infrastructure. Village is also enriched with the natural vegetation social forestry and agro-forestry programmes. The cultivable land of village Sarai Khema is 207 hectares while 234 hectares land is lying waste as banjar land, old fallow land, fallow land, current fallow, water logged land, usar land and other types of land not available for agriculture, if such land is brought under cultivations for agro forestry, social forestry plantation programmes, it will be helpful in maintaining the ecological balances of the village, maintaining the degraded environmental conditions of village and it will also be an asset to meet the increasing demand of growing cattle and human population for the fuel wood, fodder, fibre, fruits, fisheries, food grains and vegetables which will be an asset for the integrated development of the village Sarai Khema of Amethi Block, District Amethi.

2

Indigenous Knowledge for Resource Management: Perspectives of Jamuna River Chars in Bangladesh

Shaikh Shahriar Mohammad and *Sarawat Rashid*

Introduction

Main theme: Islands *(chars)* has a finite life, which are susceptible to changes in the direction of the main river channel. Hence the *chars* have a unique vulnerable to annual flooding and a limited lifespan. So impact of seasonality and resource management on the livelihood of the island *char* dwellers is very vital.

Major objectives: The study aimed to find out impact of seasonality on livelihood and to investigate the resource management for coping with the seasonal variation of the study people when they faced vulnerabilities and scarcity.

Methodology: The study was conducted in one *char* named Sthal Noahata. PRA tools were selected and used at different levels to extract qualitative information, viz. Transect walk, Social Map, Well-being analysis, Time line, Matrix, Mobility, Pie chart, FGD, Key informants interview. Beside this questionnaire method was conducted collecting the income and expenditure data. Considerable data was gathered from the secondary sources.

Major findings: Study findings indicate though season plays vital role to occur flood, scarcity, drought, heavy rain and loss of crops, using indigenous strategy *char* dwellers cope with season. They divided seasons as per their own understanding to fix up a year round activity plan of cultivation, storing and consumption of food, migration and even the taking loan.

Conclusions: The study prepared an in-depth profile of the selected *char*; and findings would help to understand the ways that *char* dwellers identify own strategy to cope with seasonal variation in the context of their own way of resource management.

Human beings have been using different components of this world in pursuit of its existence for exploring various strategies for resource management (Ackernecht, 1971). Indigenous knowledge is very important and an essential part of rural people for their resource management strategies. Indigenous knowledge provides the basis for problem-solving strategies for local communities, especially the poor (Alam *et al.*, 2000). It represents an important component of global knowledge on development issues (Amin, 2002). Indigenous knowledge is an underutilised resource in the development process. By investigating it was identified that indigenous knowledge play important role to know what the local communities use in their resource management system, it helps them to improve understanding of local conditions and provide a productive context for activities designed to help the communities (Medical Anthropology Unit, 1995). Understanding indigenous knowledge can increase responsiveness to clients (Choudhury, 2002). Adapting international practices to the local setting can help to improve the impact and sustainable development (Dunn, 1968). Sharing indigenous knowledge within and across communities can help enhance cross-cultural understanding and promote the cultural dimension of community believes (Flavier, 1995). Most importantly, investing in the exchange of indigenous knowledge and its integration into the programmes of the World Bank and its development partners can help to reduce poverty effectively (World Bank, 1991).

The people of the rural Bangladesh have traditional knowledge about resource management (Hasan, 2000). They depend on it for various reasons, such as socio-economic status, cultural pattern as well as lacking of modern facilities (Helman, 1990).

The main concept of this field research is indigenous knowledge for resource management. This indigenous resource management strategy has never been studied in broader perspective over the ages. Generation to generation this knowledge had been adopted or learnt from the predecessors (Chowdhury *et al.*, 1996). This resource management system had never gone through in a scientific way and no potential research work has been carried out yet (Karim, 2000). The main focusing point of this field research is to find out impact of seasonality on livelihood and to investigate the resource management for coping with the seasonal variation of the study people when they faced vulnerabilities and scarcity.

Objectives of the Study

The study aimed was as follows:

- To identify the coping strategy of the seasonal variation of the study area; and
- To assess the vulnerabilities and scarcity level of the island char.

Methodology

The study was conducted in one *char*, named Sthal Noahata. PRA tools were selected and used at different levels to extract qualitative information, viz. transect walk, social map, well-being analysis, time line analysis, matrix map, mobility map, FGD, key informants' interview. Beside this questionnaire method was conducted collecting the income and expenditure data. Considerable data was gathered from the secondary sources. The Data of this study was collected during 2006-07.

Limitations of the Study

This is common to a research work to face certain limitations. There are few limitations that the researcher had to face are:

- It was taken more time to conduct some PRA tools. So sometimes the participants reluctant to response properly.
- For considering char dwellers' busyness in day time, the survey team conducted their sessions at night. But it was really difficult to conduct a session at night.
- The team had the plan to collect data for last 3-4 months by every visit. But sometimes for some cases it was totally difficult to collect appropriate data in due visit.

Village Profile

Char Sthal Noahata is an island char located at Sthal union in Chauhali upazila under the Sirajganj district. It is bounded on the north by Digholkandi village, on the east by Ekrampur and Katarbari, on the south by South Teghuri and on the west by the mighty Jamuna river. The char village occupies an area of 850 acres. This village basically is a part of an island char accumulated nine villages spread over two unions. Char Sthal Noahata is divided into two parts known as South Noahata and North Noahata through a natural canal (*Khal*); but in dry season it becomes like as dried depression strip when it is easily crossed from one part to another.

History of the Char

In British period char Sthal Noahata along with surrounding villages was controlled by Hindu *Zamindars (landlord)* but after partition in 1947 they had migrated to India for ever. In that time it was a prominent village especially for its status of education. The second oldest high school of entire Sirajganj district was situated there, but 30 years before the school and the entire char was demolished by the river Jamuna. After that, char Sthal Noahata had been only found out into *Mouza* Map but not in real world until 1988. After 1988 severe flood, the char emerged in the river bed. That time basically the South Noahata was appearing continuously. But North Noahata was emerging around 1992-93. In that time the earlier char dwellers settled down in the southern part as well as starting cultivation. At present char dwellers including the people from erosion prone areas are living in this char but South Noahata is comparatively populated. Sthal Noahata is fairly new compared to other existing villages.

Physiographic Condition

As a part of island char the topography of char Sthal Noahata is almost homogenous. But the southern part is comparatively higher than the northern part and most of the settlements built up there. Some canals are found there especially north-western part. A shallow depression strip is situated at eastern side through north to south which is generally known as *'nama jamee'*. In the period of regular flood entire char is gone under water for two to three months but the low land lies surrounding the periphery that gone under water for up to five months. On the other hand in dry season scarcity of water is found even for irrigation.

Land

According to the level of river water there are two types of land, i.e. mid-land and lowland, in Char Sthal Noahata. Mid-land occupies the area of settlement and market place. Arable and fallow land cover the lowland area even sandbar also includes here. Sandy loam soil occupies the entire char except sandbar and *'nama'* (shallow depression strip or dead canal) which are covered by sandy and loamy soil respectively.

Agriculture

Excluding settlement two-thirds of the total land is under cultivation. Rice is the leading food crop in Char Sthal Noahata. On the basis of producing rice the char dwellers divide their agricultural calendar into three seasons *(mousoom)* followed as *Aman Mousoom* (Mid-August to Mid-December), *Boro Mousoom or Choitalee Mousoom* (Mid-December to Mid-April) and *Aush Mousoom* (Mid-April to Mid-August). Other crops include jute, potatoes, sweet potatoes, various pulses, onion, chilli, vegetables. Fruits (Mango, Jackfruit, Blackberry, Guava, and Jujube) and timber (Eucalyptus, Acacia, Epil Epil, *Shishu*) trees have grown surrounding settlement. Bamboo is also an important product. Cattle are numerous, rose for milk, meat and also dung (a source of fuel) especially north Noahata. Most fish are available from the river Jamuna but in wet season fish are also available in canals connected directly to river for four to five months.

Population

The estimated population of Char Sthal Noahata is 2,000. The actual population and household of South Noahata (7 December, 2006) are 1,413 and 246 respectively. On the basis of socio-economic status they are themselves divided into four groups, like as rich (basically they are not rich, we can use the term of 'well-off'), middle, poor and extreme poor.

Table 2.1: Population

Group	*Male*	*Female*	*Total*
Rich (Well-off)	85	77	**162**
Middle	158	136	**294**
Poor	359	338	**697**
Extreme Poor	128	132	**260**
Total	**730**	**683**	**1413**

Source: Field Visit.

Table 2.2: Household, Family Size and Male/Female Ratio

Group	*No. of Household*	*Family Size*	*Male/Female Ratio*
Rich (Well-off)	19	8.53	110.4
Middle	44	6.68	116.2
Poor	125	5.58	106.2
Extreme Poor	58	4.48	97.0
Total	**246**	**5.74**	**106.9**

Source: Field Visit.

Table 2.3: Number of Female Headed Households

Rich (Well-off)	0
Middle	0
Poor	1
Extreme Poor	12
Total	**13**
Female Headed Households: 5.28%	

Source: Field Visit.

Health Status

Health facilities are very poor in the present study area. This is because there is no health centre or clinic or a doctor. A local village healer (*kobiraj*) is available who is generally prescribed for illness as well as a midwife is very important charlady for the time of delivery cases. The main source of drinking water is tube well but if one is not available every house and also the sanitation system of overall not up to the mark especially this condition is measurable for the poor and extreme poor.

Table 2.4: Assets of Different Groups

Group	*No. of Households*	*Tube well*		*Latrine*		*Boat*	
		No.	*%*	*No.*	*%*	*No.*	*%*
Rich (Well-off)	19	18	94.7	13	68.4	13	68.4
Middle	44	34	77.3	15	34.1	13	29.5
Poor	125	62	49.6	32	25.6	4	3.2
Extreme Poor	58	11	19.0	15	25.9	3	5.2
Total	**246**	**125**	**50.8**	**75**	**30.5**	**33**	**13.4**

Source: Field Visit.

Education Status

Like as the health status, the education status of Char Sthal Noahata is very poor. There is only a registered primary school which does not carry out better service for the school going children. Besides it, there are also two *moktob* (primary level religious institutes) where English is not taught. The char dwellers aware about education but they can't or lose their interest to send children to high school because of distance to school, time and transport. Female pupils suffer more trouble about this matter. However, some children attend high school in the neighbouring village Teghuri even a fewer portion of them stay at Enayetpur or Chauhali in their relative's house.

Infrastructure

Char Sthal Noahata is a part of an island char, so that the infrastructural development is not found at all. A table of village infrastructures is given below:

Table 2.5: Village Infrastructures

Infrastructure	*Number*
Primary School	1
Moktob	2
Mosque	2
Market (*Bazaar*)	1
Shop	2
Loom (*Tatghor*)	12

Source: Field Visit.

Use of Indigenous Knowledge to Overcome the Food Vulnerabilities

Seasonality

After completion the social map seasonality calendar of scarcity was drawn. It was ensured that participants from each part of the village would present. Then the villagers prepared this calendar using jute stick on a brown paper. Their prepared seasonality calendar was same as Fig. 2.1.

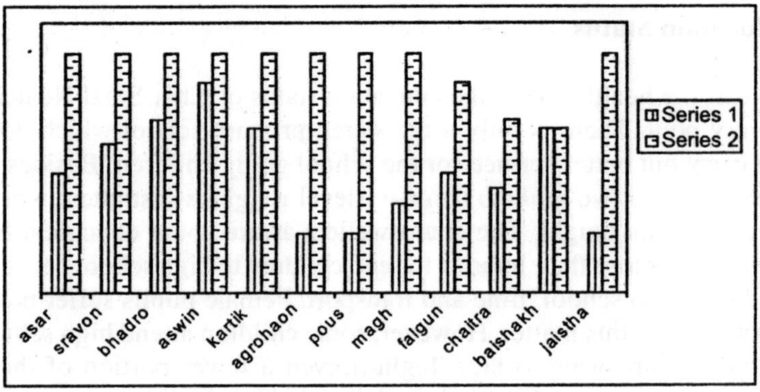

Here,
Series 1 = year of occurred a regular flood
Series 2 = year of occurred a high flood or flash flood

Notes: The participants demarcated some events occurred every year which related to poverty or scarcity. These are noted below:

Flood = *Ashar* to *Ashwin* (Mid-June to Mid-October)
Scarcity = *Joishtha* to *Magh* (Mid-May to Mid-February) {if high flood is occurred}
= Shrabon to Kartik (Mid-July to Mid-November) {if regular flood is occurred}
Felicity = *Agrahayon*, *Poush* and *Joishtha* (Mid-November to Mid-January and Mid-May to Mid-June) {if regular flood is occurred}
Break Even = *Magh* to *Choitra* (Mid-January to Mid-April) {if regular flood is occurred}
Migration = *Agrahayon* to *Choitra* (Mid-November to Mid-April) {if high flood is occurred}

Fig. 2.1: Seasonality Calendar of Scarcity

In this study seasonality is the most important tool to be undertaken. Depending on this tool time of field visit was decided. It was asked the gathering to tell us how they divide a whole year on the basis of seasons. A sheet of brown paper and colour pens to draw the picture were given. Respondents drew a line to symbolize the whole year, and then they divide it into three seasons or "Mousum".

- Firstly "Aman mousum",
- Secondly "Boro/Irri mousum", and
- Lastly "Pat/Aush mousum".

They also divided each of these mousums into four Bangla months and wrote down the names on the paper. Months covered by each mousum are as follows:

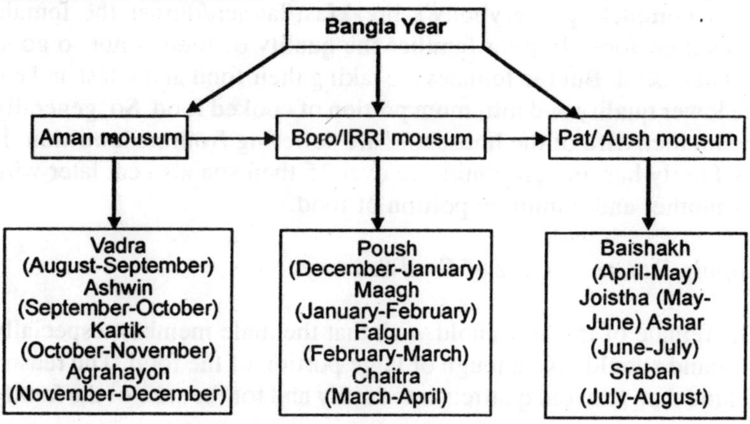

Diagram: Seasons Divided by the Villagers

Intra-Household Food Consumption Patterns

Male

They feed the male people who are earning money for family first. The reason behind:

- Male are naturally doing the hard work.
- Male do not tolerate starvation like female.
- Male food consumption ratio is higher than female.
- The lives of the wife in the hands of their husband.

- They are always hot tempered, so for intra-house peace and harmony female naturally give the large portion of food and serve first.

Children

In case of child, if the age of the child is under 10 years then the mother and also the father give them food earlier. But 10-12 years old girls grow up as her mothers' ways. She takes her food after the male member. But if the child is boy then he also served with his father.

Female

After completing everybody's breakfast/launch/dinner the female takes their food. In poor families the quality of food is not so good and balanced. But the females are taking their food at the last and eat the lower quality and minimum portion of cooked food. So, generally the mother/wife of the household are suffering from malnutrition. If the family has any girl child age over 15 then she also eat later with his mother and minimum portion of food.

Coping Pattern in Case of Food Scarcity

The female of the household wish that the male member; especially husband should take enough or large portion of the food. The reason behind that, to avoid quarrels into family and for the sake of the family harmony.

Use of Indigenous Knowledge to Manage Work Scarcity

Scarcity of Works by Seasons

Flood is an unavoidable fact of Chars' people lives. They are growing up with facing flood, severe rain, river erosion, sandy hot wind, drought etc. natural calamities and also passed their lives in vulnerabilities, migration, shifting house and so on. In char area people are mainly working as a day labourers, weavers, farmers, fishermen, rickshaw/van puller, milkman, fuel seller, shopkeepers etc. here below Table 2.1 we can see their Scarcity of Works by Seasons.

Table 2.1: Scarcity of Works by Seasons

Occupation	*January*	*February*	*March*	*April*	*May*	*June*	*July*	*August*	*September*	*October*	*November*	*December*	*Reasons Behind Scarcity*
(1)	(2)	(3)	(4)	(4)	(5)	(6)	(7)	(8)	(9)	(10)	(11)	(12)	(13)
Agri-labourer	95%	50%	33%				50%	95%	95%	95%	50%	50%	➢ Flood ➢ River erosion ➢ Drought ➢ When the crops are existing in the field (*monga period*)
Rickshaw/ van puller			20%				50%	95%	95%	95%	50%		➢ When the roads are gone to under water during flood period ➢ When the crops are existing in the field (*monga period*); because harvesting period they are busy with carrying crops to home or market.
Farmer	95%	50%	33%				50%	95%	95%	95%	50%	50%	➢ Flood ➢ Drought ➢ When the crops are still in the field (*monga period*) ➢ Agricultural inputs get higher price in the market. That time are somehow managing their inputs. Even selling out their valuable assets, like; cow, stocked rice for the rest of the months, ornaments etc.

(*Contd.*)

Table 2.1 (*Contd.*)

(1)	(2)	(3)	(4)	(4)	(5)	(6)	(7)	(8)	(9)	(10)	(11)	(12)	(13)
													➢ Takes loan for buying agricultural inputs from the NGOs, moneylenders, local village associations (they give credit in higher interest rate), relatives etc. sources.
Fisherman								50%	80%	80%	80%	50%	➢ Flood period, because the picicultured are flash out and almost all are engaged in fishing. So, the fishes are available in markets. The price is fallen. ➢ In the rainy season they do not go for sell out their caught/collected fish to the Enayetpur Bazaar.
Milkman							33%	66%	80%	80%	33%		1. Many milching cows are died due to flood. 2. During flood period the total milk production is decreased due to scattedness of cows (usually send to embankment or relatives' house) and also its owner. 3. Communication/transportation system is collapse so they cannot collect the milk from the owner's house. 4. The price of milk falls due to proper marketing channels. 5. Supply of milk is much lower than the demand in local market. 6. Production of milk falls when the cows do not get enough food and fodder at the rainy season/flood period.

(*Contd.*)

Table 2.1 (*Contd.*)

(1)	(2)	(3)	(4)	(4)	(5)	(6)	(7)	(8)	(9)	(10)	(11)	(12)	(13)
Shopkeeper							20%	50%	50%	50%	20%		➢ Flood ➢ River erosion ➢ In the crisis period of others, because the total sales is fallen. ➢ If price of goods is hike up, because total sell out is fallen and net gross profit is lessened.
Fuel seller							50%	95%	95%	95%	33%		➢ Flood ➢ Rainy days; they do not collect the cooking fuel and their stock of fuel is damaged. ➢ Crisis period of others, because the total sales of cooking fuel is drastically fallen.
Weaver							20%	50%	80%	33%	10%		➢ Flood ➢ Rainy days ➢ Lean period (*monga period*), because the total sell out of cloths is fallen. ➢ Before the festivals, because they have to invest a lot for production purpose.
Boatman			33%	66%	66%	33%		33%	66%	50%			➢ Drought ➢ Heavy rainy days, people do not want to go outside the home. ➢ Lean period (*monga period*), the people who are across the river, do not pay *monga period.* They have to wait till the crisis period.

(*Contd.*)

Table 2.1 (*Contd.*)

(1)	(2)	(3)	(4)	(4)	(5)	(6)	(7)	(8)	(9)	(10)	(11)	(12)	(13)
Agri-business				20%	20%			50%	80%	80%	33%		➢ Flood; crops are damaged. ➢ Rainy days, they do not sell out much. ➢ Drought, the production is fallen. ➢ If suddenly fall in the price of their stocks, goods/crops. Because, they bought it on higher price and invest a lot.
Cow broker							50%	95%	95%	50%			➢ Flood period, because the weekly cow bazaar (*cow haat*) is not sitting. ➢ They engaged more in the flood period with their own family.

Coping with the Seasons (Round the Year)

In Sthal Noahata the people take several steps to cope with various seasons in many ways. People are usually:

In the Rainy Season/Flood (Mid June-Mid September)

- Store rice, *muri* (rice crispy), *chira* and keep this in safer place.
- Made an extra *chula* (burner); which is mobile.
- Keep some fine firewood.
- Made a temporary storage place high above the flood level. Even, sometimes they made it at the top of the house using bamboo, jute fibre, eucalyptus trees, strong branches of trees etc.
- Keep few cash in their hands.
- Made floating platform using banana trees. This is also used as boat for the purpose of moving from one place to another.
- If they have tube-well then they raise it up to the flood level. If not possible, then collect the drinking water from the safe tube-well. Some cases, for collecting drinking water they have to travel a lot.
- If they have any valuable things, then they send it to the relatives' house in safer place which area does not affected by the flood. But if it is not possible then they hide these things in a safer place or keep it with his bags or *sari's anchol* (end part of the female's long wearing cloths).
- For the defecation purpose, if there is no toilet above from the flood level, both male and female go by the *vaala* (banana made floating platform) nearby bushy area or away from the people.
- In the flood time, they are tried to avoid taking bath. But if necessary and do not manage to collect tube-well water takes their bath in the flood water.
- Washing their cloths in the flood water.
- When the rain is continued few days and they guessed that the flood is occurred then they send their livestocks to the nearest embankment (*Monacosha* embankment) or relative's house, who are free from the flood.
- In case of goats and poultry birds, they are just sold it out

immediately. Because, in the flood, it is difficult to manage them and very often they are stolen, affected by diseases and dead by the snake bite. So, they sold it and try to overcome their immediate problem or needs with this money.

- During the flood, those have tried to save cattle, if they have not in big trouble do not sold. They tried by all means to save it. Because, if they try to sell cattle in this time, the buyers willing to pay just half of the actual price. So, it is a great loss for them.
- They send their cattle with others. Generally, 5 to 6 cattle are looked after by one cowboy or member. So, they send cattle and sometimes male family members together.
- During the flood period, the weaver of the Sthal Noahata falls in deep trouble. They have the mechanism to raise weaving machine up to one foot. So, if the flood water comes into the weaving sheds they can continue their works.
- If the severity of the flood is higher, if possible, they send young boys and girls to their relatives' house situated in the flood free area.
- To cope with the food scarcity during the flood, they usually take their meal one or two instead of three times.
- During the flood, there is a scarcity of vegetables and the prices hike up. So, they use more chilies (both green and red) to avoid using excessive vegetables. The reason behind that, if the dishes of vegetables is hot-tasting then it takes minimum. In this way, they can save some money instead of buying vegetables.
- They take the green banana, the inner portion of banana tree, the flower of banana tree, lotus and water born vegetables which are available at low cost or free of cost.
- During the flood, all low ditches and ponds are flash out. So, the fishes are freed and scattered. The family members usually tried to catch fishes and take fishes very often with meals. This time availability of fishes is increased in the market and prices go down.
- Tried to manage VGD, VGF cards from the Union Parisad.
- The rich people, those yearly consume food from their own production, sell out their excess rice to overcome recent hazards.

- The poor people get loan from the moneylender, NGOs, local village association.
- Try to get relief from the voluntary groups or governments.
- All family members are trying to get works. Even, those family members who have never done odd/hard works.
- Borrowed food stuff from the nearest and known grocery shops.
- Women of the house naturally sowing seeds of various vegetables, without uprooting the earlier trees. When the new vegetable trees like, pumpkin, bottle-gourd, bean etc. are matured then they uprooted the old trees.

In the Lean Period/Monga Period (Mid September-Mid November)

- Borrowed food and cash from the relatives.
- Takes loan from the NGOs, local village *samety* (association), and moneylender.
- Male members of the family migrated to Sirajganj, Rajshahi, Bogra and Dhaka as a day labourer.
- For cash and buy food sell out their asset, if they have any.
- Send their children to other's house; especially city areas; as home servants.
- Take the foodstuff which is cheaper.
- Try to gather vegetables from the roadside or fellow lands especially the inner side of banana trees.
- Take foods once or twice a day instead of three meals.
- They try to take cooked food once a day and another is just dry food like, *muri* (rice crispy), *chira, and fried coarse rice etc.*
- Take little amount of food as a meal.
- If they are deep-in-trouble and does not get any work then begging for the sake of their children.
- And last but not the least, if all means of collecting food is failed, than they have nothing to do with live in hunger.

In the Winter (Mid November-Mid March)

- The winter season is comparatively comfortable season for them.
- This time they get available works. So, they try to keep some

savings for their future needs and if they have any loan then try to repay.

- In this season they take the various types of vegetables and sufficient foodstuff, to keep their body sound.
- Those who have the arable land they get the new crops (*Aman rice*), which makes them solvent.
- Those who have no land (landless/extreme poor) they also get money from working as an agricultural labourer or day labourer of the well-off group's crop field.
- Those who are extremely poor, if they have not enough warm cloths; try to keep their body warm by making a fire and just sit besides.
- In this time, fishes are not available, so the price of fishes are higher. They cope with the situation and overcome the protein deficiency by taking eggs comparatively more.

> The most interesting feature of migration is, the male members temporarily migrate to the city or adjacent district, where they might get work, in their well-off period. The reasons behind that, in this period of time they have some money in their hands, so they give this amount of money to his wife's hand for maintaining the family and the male member, usually the husband, migrate for two to eight weeks. In the meanwhile, the wife maintains the family and tried to save some money from working in other's houses or crop field. Even, working as a labourer of earthen works. So, when the husband come back to the house, he can maintain the family from one to three months (actually it depends on how many days he worked and how much they saved) with the saved money.

During Summer (Mid March-Mid June)

- They can mange some agricultural works.
- The price of fish is higher, so they try to take eggs and pulses to overcome the protein deficiency.
- In the summer, the weavers of Sthal Noahata are usually working more than the rest of the year. Because, in Noahata there is no electricity. So, the weavers can work more in the day-light.
- The rickshaw/van puller and boatman both are engaged little bit with carrying out the *dhoncha*, black gram, rice etc.
- Women are sowing the seedling at their homestead and get sufficient vegetables. One of the respondents said that, they are hardly ever buy vegetables in cash. It is saved a lot of money.

- If they have cash in their hand, they usually buy the productive assets, like poultry birds, cattle, goat etc. It is not only their means of immediate needs but also increase their income.
- In char lands, there is hot strong wind blows occasionally, it causes sometimes catch fire in the house. During hot windy period, char dwellers cooked their food in the morning, because, after late morning the hot wind blows.

Taking Loan as a Means to Cope with Scarcity

The reasons behind why the char people take the loans from different sources are many. They usually take loans for their crisis period. The Sthal Noahata people generally do not go for the loan. Table 2.2 shows the priority/rank from whom they borrowed in cash or kind.

Key Findings

Reasons for Changing the Socio-Economic Condition

There are many reasons for changing the socio-economic condition or status. Especially in the char area the inhabitants are the most vulnerable group of the people. There is a class in the Sthal Noahata char regarding their socio-economic status (see Fig. 2.3).

In the char areas we are found some information for changing their socio-economic condition. The change is both upward and downward. Those who are loosing their property/status with stable situation we mentioned them *declining family* and those who are virtually increased their well-being we mentioned them *improving family*.

Reason for Declining

- The earner son now lives in separate, so the earning of the family is declining. Because the parents are now aged. They do not work hard.
- Loss of land by river erosion.
- Land acquired by government for the purpose of cluster village.

Table 2.2: The Priority/Rank from whom People Borrowed

Ranking/ Priority	*Name of the Source*	*Reason for taking loan and means of loan*
(1)	(2)	(3)
1	Neighbour	*Advantage:* ➢ Give loan cooked food ➢ Cereal food ➢ Scanty amount of money ➢ Do not pay any interest ➢ Take loan for a shorter time ➢ Fulfil the immediate need ➢ Any time tried to help them *Shortcomings:* ➢ Repayment the loan within the given time. ➢ If not may insult and quarrel with them. ➢ If serious, then they must pay it by taking loan from the NGOs, like MMS, BRAC, Village Association with high interest rate.
2	Local grocery shop	*Advantage:* ➢ All kinds of grocery items. ➢ Borrowing any time of the months. ➢ Always give loan in kind (not in cash) with the limit of certain amount of Taka. *Shortcomings:* ➢ They must pay the money as a whole. ➢ The date is generally 1st date of Bangla New Year (Boishak = mid of May). The shopkeeper arranges a programme of collecting all his money from the borrower. This festival called *halkhata* means end of current leaser book and open new book for new Bangla year. ➢ Never give loan in cash.

(Contd.)

Table 2.2 (*Contd.*)

(1)	(2)	(3)
3	Local village association	*Advantage:* ➢ Immediate need; when they cannot manage from the relatives or neighbour. ➢ Hardship. ➢ Fertilization of the growing rice field. ➢ Sudden need for treatment. ➢ Buying seeds, fertilizers. ➢ All other emergency need. *Shortcomings:* ➢ The interest rate is highest. ➢ Pay the installment weekly.
4	Moneylender	*Advantage:* ➢ Taking loan in the rainy season. ➢ In the period of flood. ➢ If they have not any work. ➢ There is scope to repay the loan as working in the moneylender's crop-field. ➢ They not only give loan in cash but also in kind, like Rice etc. *Shortcomings:* ➢ The interest rate is higher. ➢ If they take loan from the moneylender they are usually bound to obey their direction/decision
5	Relatives	*Advantage:* ➢ Give loan if it is within their capacity. ➢ Always try their best to help them. ➢ Give loan when they buy a valuable things. ➢ In case of their treatment. ➢ In case of the wedding ceremony. ➢ Never claim any interest. ➢ Even, sometimes they just waiver their loan.

(*Contd.*)

Table 2.2 (*Contd.*)

(1)	(2)	(3)
		Shortcomings: ➢ If they do not repay money at the given time; it may occur a great family problem. ➢ If the loan received from the wife's fathers' house then the husband try to avoid repayment. If the wife's father/brothers claim the money then sometimes it causes the domestic violence; wife is bitten by the husband. ➢ Sometimes, the relatives unwilling to give loan. Because, they feel afraid that they might not repay this money and do not want to show the relatives that they have the enough money for giving loan to others; it may encourage other relatives.
6	Local NGO (MMS)	*Advantage*: ➢ Small business purpose. ➢ Agricultural purpose. ➢ Buying cow. ➢ Treatment purpose. ➢ In case of scarcity of money. ➢ Local level NGOs, so the management process is easier. ➢ In case of marriage purpose. ➢ If the amount of money is higher. *Shortcomings*: ➢ They collect the installment weekly, from the very next week. ➢ The rate of interest is higher (16% -20%).

(*Contd.*)

Table 2.2 (*Contd.*)

(1)	(2)	(3)
7	Local BRAC office	*Advantage*: ➢ Small business entrepreneurship. ➢ Agricultural purpose. ➢ Buying cow. ➢ Treatment of disease. ➢ If they fall in severe crisis. ➢ For the marriage of daughter. ➢ If the amount of money is large. *Shortcomings*: ➢ The repayment of installment is weekly. ➢ The condition of the loan is tuff.

In the char areas the admirable fact is that there is no perpetual well-off group. There is always the risk behind the well-off people. The considering facts and factor are:

- ➢ The Chars are emerged for a short time span.
- ➢ In the Jamuna river chars are existing usually 6-14 years.
- ➢ The landlord of the chars became landless within few years, in some cases in a year by the river erosion.
- ➢ Even though, the landlord became poor people without loss his land. If the annual floods carry the sand into his fertile land, then the land becomes useless.

- ➢ Abdullah Munshi, the rich person in this village, stated that, 'In the eyes of nature, here all are equal none are more equal than others'.
- ➢ Motaleb a milkman of this village said that, 'In Char areas usually they give priority cash money holder, owner of cattle, permanent job holder, those have link with city areas and capacity to migrate other places with work, instead of landholder of char areas.'

Fig. 2.3

- Stepbrother takes all arable land. He has not any arable land. He now lives in a hired land of others.
- After father's death, the earning is declining.
- In an accident earner's hand was broken. Though get a lot of treatment but his hands are not curing. So, he does not work and lost his earnings.
- Having one or more than one daughter. So, when he arranges marriage for them, these families put in deep trouble and declining as a poor.
- Faced a severe accident, which cost a lot of money. For that reason he sold his land and now declining.
- Sold-out the valuable asset, which was the main income source of the family, like boat, horse and cart, milking cow etc.

Reason for Improving

There are various reasons for improving condition and became a well-off. These are:

- Mortgage others cattle and land.
- Have grown skills and get works frequently.
- Some land is arisen from the newly emerged char.
- The boys of the family are now engaged in work and earns.
- Have cattle, so the cow gives birth to calves. From it the family is improving their economic condition.
- Mortgage others land and later buy the land.
- To grabbing the *Khas land* (government land/unclaimed land).
- Both works as day labourer and also taken *Adhi* (mortgage) the cattle of others.
- Children go to school. So, wish that after few years they get the returns. Here they considered their school going children as asset.
- Take a piece of land as a dowry. After that mortgage others cattle. In this way, changing economic condition.
- By selling grocery items. Earns cash and lend the money with interest.
- Takes cash money as a dowry. Separated from the fathers' household. Members of the household are fewer than previously.

Conclusion

The study prepared an in-depth profile of the selected *char* Sthal Noahata. findings would help to understand the ways that *char* dwellers identify own strategy to cope with seasonal variation in the context of their own way of resource management. In study area it was found that the use of indigenous knowledge for Ecological resource management is socially and culturally embedded and it is dynamic. This knowledge is learned by themselves through their practical experiences and very often it is transmitted from one generation to the next through unwritten and oral communication. Some organization; such as—World Bank, World Health Organization, United Nations Educational, Scientific and Cultural Organization, Food and Agricultural Organization are giving importance to it. But from the context of our country, a huge number of researches on indigenous resource management strategies need to be done. And all these research works should be conducted with the participation of the indigenous people. This research is such an endeavour to formulate policies for conserving and restoring the local people's skills and indigenous practices.

REFERENCES

Ackerknecht, Erwin H. 1971: *Medicine and Ethnology: Selected Essays.* Baltimore: Johns Hopkins University Press.

Alam, M.K. and Khisa, S.K. 2000: The Perception of Ethno Botany in Chittagong Hill Tract and Its Linkage with Biodiversity. In: Alam, N.A. (ed.), *Of Popular Wisdom: Indigenous Knowledge and Practices in Bangladesh*, August 2000, BARCIK, Dhaka.

Amin, M.N. 2000: Potential use of Indigenous Knowledge in Sustainable Conservation of Plant Diversity in Bangladesh. In: Alam, N.A. (ed.), *Of Popular Wisdom: Indigenous Knowledge and Practices in Bangladesh*, August 2000, BARCIK, Dhaka.

Choudhury, A.F.H. 2000: Indigenous Knowledge System: A Development Paradigm in Anthropology In: Alam, N.A. (ed.), *Of Popular Wisdom: Indigenous Knowledge and Practices in Bangladesh*, August 2000, BARCIK, Dhaka.

Chowdhury, J.U., Alam, M.K. and Hassan, M.A. 1996: Some Traditional Folk Formularies against Dysentery and Diarrhoea in Bangladesh, *Journal of Econ. Tax. Bot.*, Additional Series-12.

Dunn, Frederick L. 1968: *Traditional Asian Medicine and Cosmopolitan Medicine as Adaptive System* (*Imperialism and Medicine in Bengal*—Poonam Bala).

Flavier, J.M. *et al.* 1995: The regional programme for the promotion of indigenous knowledge in Asia. In: Warren, D.M., L.J. Slikkerveer and D. Brokensha (ed.), *The Cultural Dimension of Development: Indigenous Knowledge Systems*, London.

Hasan, S. 2000: Indigenous Perceptions, Predictions and Survival Strategies Concerning Cyclones in Bangladesh. In: Alam, N.A. (ed.), *Of Popular Wisdom: Indigenous Knowledge and Practices in Bangladesh*, August 2000, BARCIK, Dhaka.

Helman, C.G. 1990: *Culture, Health and Illness*, London.

Karim, A.H.M.Z. 2000: The Indigenous Pattern of Subsistence Practices of the Munda and Mahato People in the Sundarba Mangrove Forest Areas of Bangladesh: An Anthropological Overview. In; Alam, N.A., ed., *Of*

Popular Wisdom: Indigenous Knowledge and Practices in Bangladesh, August 2000, BARCIK, Dhaka.

Medical Anthropology Unit (ed.) 1995: *Applied Health Research Manual: Anthropology of Health and Health Care*, University of Amsterdam.

The World Bank, 1991: *Using Indigenous Knowledge in Agricultural Development; World Bank Discussion*, By Warren, D.M. Washington, D.C.

3

The Current Perception of the Forest Juju in Ecological Conservation in an African Context

Humphrey NGALA NDI

Introduction

The Collins English dictionary defines a *juju* as an object superstitiously revered by certain West African peoples and used as a charm or fetish. In fact, daily life in most of the traditional African society is greatly regimented by the belief in all sorts of *jujus* and their abilities to inflict various punishments on people who violate societal norms. Man-Environment relationships especially the exploitation of the soil, forest, water, and hunting amongst others are equally controlled by the *juju* in these communities. Although the influence of the belief in *jujus* is fast waning in Africa because of the influence of Westernisation and rapid urbanisation, there is equally a strong feeling amongst development workers including conservationists that African development should be inspired more by its knowledge systems than by those adopted from outside. It is in this string of thought that Indigenous Knowledge (IK) (2004) notes the conservation abilities of the Barotseland people in the Upper Zambezi where it was a serious offence to kill a female animal, and hunting was prohibited during the breeding season. Harvesting food and fuel wood in the montane and gallery forests of the *Mbumland,* Bamenda plateau of Cameroon was regulated with the scare of various *jujus* roving the forests in the dry season and peak harvesting periods. In corroboration, Mkapa (2004), former president of Tanzania observed that as scientists struggle to respond to global challenges, they have increasingly distanced themselves from local ways of solving problems. As he puts it, local knowledge can constitute important ways of solving global problems. Sacred practices are in general, part of what is loosely described as indigenous knowledge.

Before proceeding, it is necessary to define a few terms central to this discussion:

- indigenous people;
- indigenous knowledge; and
- unclassified forests.

Melchias (2001, cited in Eyong, 2007) defines indigenous people as a culturally distinct ethnic group with a different identity form the national society, and who derive their existence from local resources and are politically near inexistent. Within the framework of such a definition, and the complex nature of social and economic interrelations within the global economy, it seems that purely indigenous peoples are near non-existent. From this definition, it can be gathered that indigenous people are marginalised and facing extinction. Some examples include the pygmies in the Congo forest and south-east Cameroon, the Tuaregs in Mali, the San people in Botswana, the Indians in the USA, the Maoris in New Zealand, and many others around the world. The UNO (1993) estimated over 300 million indigenous people living in more than 70 countries around the world, from the Arctic to the Amazon, from Asia and Africa to Australia and New Zealand. Their physical and intellectual existence is threatened by the ever expanding frontiers of modern civilisation into remote areas notably forests for farmland, energy, timber and minerals.

Indigenous knowledge on its part can be defined as a set of perceptions, information, and behaviours that guide local community members' uses of land and natural resources. Indigenous knowledge is created and sustained by local community members as a means to meet their needs for food, shelter, health, spirituality, and savings. Sacred forests or groves constitute a crucible where indigenous knowledge is made manifest. These are special forest areas believed to harbour with supernatural powers and commonly comprising sites for rituals, initiations and other special functions and ceremonies. They occur everywhere in the world but are more significant in preindustrial societies, and areas that escaped European expansion and Christianity.

Few forests in the Western Highlands are permanent. The majority of forests stand in this region are non-permanent and therefore, unclassified either as community or communal forests. Within the context of the law on forest management in Cameroon, local populations have customary rights to harvest all forests, wildlife and fisheries products freely for their personal uses except for protected species.

Forest is central to the quality of climate, soil, water, and even game which serves as a source of protein for village communities, and a touristic resource which can be an invaluable revenue source. With the enormous cost involved in globally mitigating the current wave of environmental changes (climate) in the world, indigenous knowledge offers a great opportunity. Since lots of forests in Africa are under the custody of some traditional authority (community forests), reviving and encouraging IK in ecological conservation in a multitude of local settings could have the ability to orchestrate regional or continental wide improvements in biodiversity.

In the contemporary survey of environmental issues affecting man, the causes are well understood and wonderful remedies can be suggested. For example, deforestation is known to encourage desertification and the planting of trees in affected areas is a sure remedy; the emission of greenhouse gases is known to cause global warming and the limitation of these emissions would definitely reduce global temperature change; farming in water catchment areas reduces the groundwater store in such areas but the prohibition of farming and the planting of water loving trees can help to recharge the groundwater store... The question to ask is why environmental issues continue to preoccupy mankind notwithstanding the existing knowledge of causes and remedies some of which examples are cited above.

Therefore, the problem in addressing ecological issues today is in making the remedies socially, culturally and politically acceptable to the people. For example, agronomists know how to control soil erosion, but not how to convince peasant farmers from bush burning as a form of land preparation; engineers know how to multiply the yields of particular crops many-folds, but not how to convince governments, civil society organisations and other interest groups to accept genetically modified organisms (GMOs); scientists know how to reduce the emission of greenhouse gases but not how to convince the major polluters (U.S.A., China, India, U.K., Brazil) and others to limit their emissions. Indigenous conservation strategies seem to succeed in this area because they are constituted of socially and culturally accepted practices passed down from one generation to another by a group of people, different from the externally motivated and physically coercing strategies of governments, and their foreign partners.

This paper sets out to examine the various traditional practices in ecological management in selected localities in the Western Highlands of Cameroon, arguably the richest area in traditional knowledge systems in Cameroon. It also examines the motivation for the conservation of IK, its successes, and the threats to its survival in a globalisation friendly context. That is why this paper also ponders on the feasibility of IK making significant contributions to conservation today in the wake of the crushing influences of urbanisation, modernisation and globalisation.

The design for this research was essentially qualitative. There was an extensive survey of existing secondary material on indigenous knowledge. In addition, the paper was also informed by empirical and observation data.

The results are the documentation of traditional knowledge in ecological management in the Western Highlands of Cameroon. In the wake of several failed modern conservation strategies, it seems that policy makers are rekindling interest in the role IK can play in resource conservation. Policy makers require such knowledge to inform their decisions in community resource management.

Context of the Study

The three past decades have been marked by a generalised drive to raise awareness to the threats man was posing to his environment, the very basis of modern civilisation. As part of this drive, Lester Brown, an American environmentalist helped pioneer the concept of sustainable development in the mid-1970s. In 1987, the Brundtland Commission popularized the concept by defining sustainable development as that endeavour, which meets the needs of the present generation, without compromising the ability of future generations to meet theirs (WCED, 1987). The hopes raised by the proponents of the concept of sustainable development (Rio and Johannesburg

Summits, of 1992 and 2002 respectively) have waned considerably because the concept is not a law and cannot directly be enforceable in the context of international environmental legal systems. That is why documents such as the *Rio Declaration* are often referred to as *soft law,* or guidelines, policy declarations, or codes of conduct that set standards of demeanour but are not directly enforceable.

As a result of this state of affairs, faith has been rekindled in the potential of indigenous knowledge and traditional social regulatory systems to contribute significantly to ecological conservation. This was reflected at the first Global Knowledge Conference in Toronto in 1997, where political leaders and civil society activists from the developing countries called on the World Bank not only to be a provider of knowledge gained in its many years of experience, but to equally learn from the practices of communities so as to leverage the best in global and local knowledge systems (World Bank, 2004). In 1998, and in response to that call, the Bank launched the Indigenous Knowledge for Development Programme to foster learning from community based knowledge system, development practices in a bid to integrate them into Bank-supported programmes (World Bank, 2004). A feature of this shift has been the publication by the Bank of a monthly periodical called *IK Notes*, the initiation of the UNESCO Local and Indigenous Knowledge Systems project (LINKS) and even the consideration of the protection of indigenous knowledge from illegal patenting by the World Trade Organisation.

This paper is an added voice in support of the incorporation of IK practices into conservation strategies given the dismal performances of modern and dominantly Eurocentric models of ecological conservation which usually set aside the resource for conservation from the people. This performance is glaring in the state of the so called protected areas in Cameroon. The emphasis in this paper shall be on community forests, the domain of village communities as spelled out in the 1994 forestry, wildlife and fisheries law in Cameroon. The forest in the Western Highlands where most examples are taken is mainly deciduous.

Indigenous Knowledge Practices in Ecological Conservation

Sacred or spiritual practices underpin ecological or biodiversity conservation in the indigenous context. Rather than employing active policing strategies which depend on legal frameworks and the use of a number of uniform officers to enforce the law, indigenous strategies of nature conservation are enforced by the members of the community (Bhgwat and Rutte, 2006). Their inclination towards biodiversity conservation is enforced by the strengths and reverence they have for their spiritual beliefs, social rules and norms.

Invariably, indigenous community efforts at biodiversity conservation have tended to be limited to fauna and flora. Little or no consideration is accorded to the inorganic components of the biosphere like soil and water. There is no gainsaying the deep knowledge indigenous people have of their ecosystems and the factors that may jeopardise their ability to continue to function harmoniously in its natural setting. That is why all indigenous groups of people have a set of checks on the possible excesses that community members may exert on their ecosystems. The checks are strikingly similar in intent and content across different communities of indigenous people.

General Practices for Biodiversity Conservation

Indigenous communities often oppose the uncontrolled felling of trees and the hunting game in their groves. Rather, community members have the right to collect only dead wood for fuel. Mature trees are felled with the expressed permission of the traditional custodian(s) of the forest grove (Bhgwat and Rutte, 2006). In Cameroon, most sacred groves are mountain forests found in the dominantly savannah Western Highland region of the country. The strength of the indigenous conservation drive in these forests may depend on the fact that forests are a scarce ecological resource which may not easily regain its equilibrium if it is lost because of the seasonal nature of rainfall unlike the rainforests.

In this region of Cameroon, different village communities often observe weekly rest days like the Jewish Sabbath. This day is set aside for worship in the shrines, cultural manifestations, and visitation amongst others. On such days, nobody is allowed to get into the forest for any reason whatsoever. This is more so because the religious shrines and burial grounds for notables are located in the forests where libations and other sacrifices to the gods through the intercession of the ancestors is done. In general, exploitation is limited mainly to the rainy season and the forest is under embargo for most of the dry season.

Just as indicated earlier, hunting game is largely prohibited. There is a season set-aside for hunting game and intruders or non-community members are systematically excluded from all forest resource exploitation.

An interesting instrument for controlling forest exploitation is the use of the injunction. Injunctions do not only apply to community forest stands but also to privately owned groves especially raffia. The raffia palm bush remains a great asset offering an assortment of products for community members. From it comes wine, bags, clothes, furniture, brooms, and baskets amongst others. Injunctions, in the form of a sacred staff, often planted at the main entrance of a forest grove or raffia palm bush lasts for as long as the authorities think the resource is under threat.

Another practice encouraged in IK for nature conservation is decorum. Community members are required to comport themselves with decorum *vis-à-vis* sacred groves. That involves amongst others, avoiding desecrating sacred places with urine and faeces for example.

Transgressions against the established order are not uncommon in indigenous resource conservation. There are predominantly addressed by various degrees of punishments depending on the misdemeanour committed. The punishments are believed to come from the spirit world controlled by ancestors who are buried in the forests. There is the strong belief that their spirits are still residing and hovering therein.

There are two types of misdemeanours in indigenous ecological conservation systems: accidental and deliberate misdemeanours. In the case of the former, priests, or custodians of traditional religions may intercede to the spirits on behalf of the transgressors when a fine is paid. In the latter, the transgressor prepares to face the wrath of the spirits and so, does not intend to appease them. Generally such transgressors may be befallen by strange incurable diseases which eventually kill them. The fear of what the spirits are capable of doing, the disgrace that comes with it, and the heavy fines transgressors pay (goats and fowls notably) push people to avoid falling into disfavour with them. Even when a member of the household breaks a traditional norm, the parents need to pay fines to the priest who appeases the spirits on behalf of the member.

1. Changing Perception of Indigenous Knowledge in Africa (Micro Scale)

African perceptions of indigenous practices in ecological conservation have recently come under modification from Western education, rapid urbanisation and urbanism and Christianity.

- Many African countries now have compulsory and free primary educational programmes for their populations. Since education in many of these countries is still highly Eurocentric in conception, the new values children learn in schools often largely contradict those held by the custodians of indigenous knowledge. For example, it is becoming increasingly difficult to find successors to the priests who officiate at many traditional shrines across Africa. Educated children increasingly see traditional knowledge as primitive, regressive, intolerant and are lured by the hopes of higher education which may earn them a more respectable job and position in society.
- The cutthroat rate of urbanisation in Africa has not augured well for African traditional knowledge systems. The villages which host shrines and sacred groves are loosing population rapidly. Some traditional rulers who are custodians of traditional knowledge even reside permanently in towns. In addition, the priest, the village chief or *Fon* has lost a lot of the reverence people reserved for him in the past. This is because many villagers who migrated to towns have become rich and are able to use money to secure a more respectable position amongst their kin and kith than even the chief. They are even capable of transgressing on community norms, societal rules and spiritual beliefs and making amends with money. The village chief and priest have thus been lured by wealth to disregard community values, all to the disfavour of the strength of indigenous knowledge systems. The 2010 demographic survey for Cameroon puts its urban population at 52 per cent (BUCREP, 2010), up from just 20 per cent at independence and it is estimated to reach 80 per cent by 2030. This trend further jeopardises the future of indigenous knowledge systems.
- Christian beliefs are diametrically opposed to a myriad of

traditional African beliefs and practices. For examples, shrines are considered by Christian to constitute idol worship which the Bible condemns vehemently. The performance of cleansing rites is also regarded to as ancestral worship. Unfortunately for IK, Christianity has taken a very firm grip not only on a large majority of Africans but also in the closed village communities. Some African traditionalists actually worship Jesus, the founder of Christianity and at the same time, traditional deities. This in no small way debases their authority in the area of competence they profess in village communities.

- The situation has further been compounded by the increasing nearness of modern administration which has made the traditional priest or chief only a symbol rather than the true substance of authority. It is not uncommon for chiefs or priests to serve prison terms from cases with their subjects. Many villagers think that the ubiquity of modern administration has emancipated them from the authoritative hands of the custodians of traditional knowledge.
- The rampant loss of livelihoods tagged to traditionally export oriented crops like coffee, cotton, and cocoa dealt a severe blow to village economies in Africa. In many countries where this scenario was experienced, poverty levels soared unprecedentedly. The recent shift to food cropping in the midst of population pressure has pushed the population to hitherto no-go-zones (sacred forests and other protected areas). On the Western Highlands of Cameroon, for example, deforestation is on a steady rise. At present, almost half of the Bamboutos mountain forest is occupied by farmers and graziers. The same trend is occurring on the Nkambe plateau of the North West region of Cameroon. Figure 3.1 illustrates the extent to which human activity has contributed to the retreat of the once dense and luxuriant Njiseng mountain forest grove located just a few kilometres from the chief divisional town of Nkambe.

This forest which is also a watershed for the villages of Binka, Binshua and Tabenken has been exploited for game, fuel wood, medicine and farming. It is estimated that just about a third of the forest is left as Figures 3.1 and 3.2 show. The myths of the forest

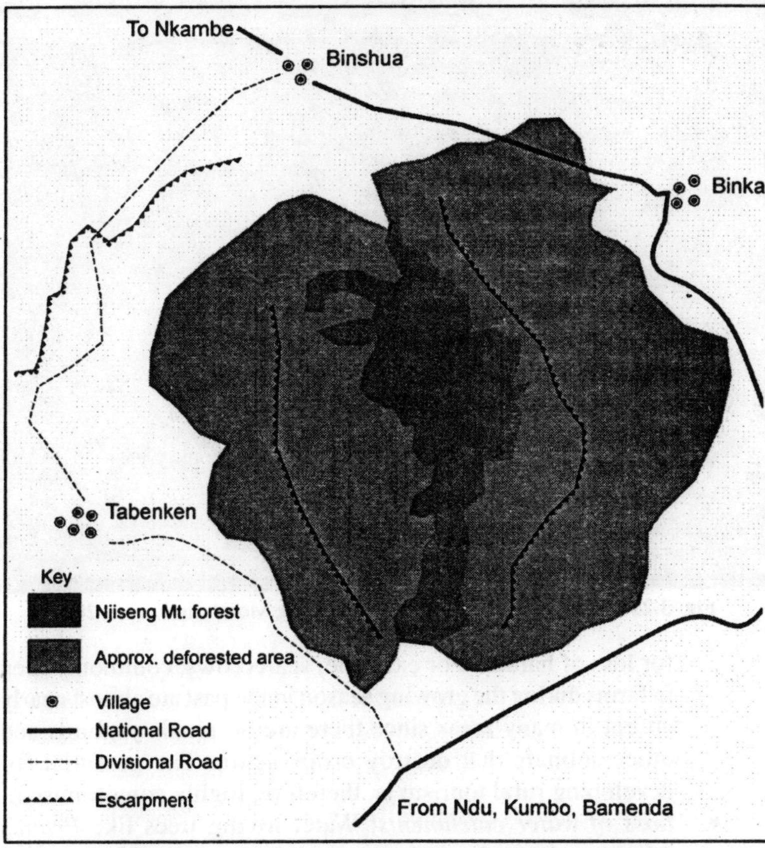

Fig. 3.1: The Remnants of the Njiseng Mountain Forest

harbouring *jujus and a forest god* have waned firstly because their habitat has receded tremendously and secondly, because of the threats discussed earlier.

2. Consequences on Biodiversity

The consequences of deforestation on rural livelihood are glaring and include:

- *Loss of rare plants and animal species*: Many species of animals that lived in almost all important mountain forests in the Western Highlands of Cameroon have been displaced

Fig. 3.2: A Google Earth View of the Njiseng Mountain Forest, 2010

for loss of habitat. For example, scarecrows, commonly seen in farms during the growing season in the past are almost nearly absent in many areas since there are no monkeys, birds and other animals that destroy crops again. The potential for developing rural tourism is, therefore, highly compromised.

- *Loss of water catchments*: Water loving trees like *Prunus Africana* have been harvested for use in modern and traditional pharmacopoeia. Raffia which has a high water conserving capacity has been over-exploited for its multiplicity of uses thereby jeopardising the water supply potential in many localities.
- *Loss of traditional knowledge*: With rapid spread of universal education, village communities in Africa are fast loosing the custodians of traditional knowledge because as mentioned earlier, the perpetuation of indigenous knowledge is becoming pretty difficult.

Discussion and Conclusion

The deep transformations taking place on the African socio-economic landscape indicate that indigenous knowledge in Africa is under severe threat from modernisation and globalisation tendencies. This has been reported from all over Africa. The beliefs ordinary Africans held in traditional religions and spirits have waned considerably because of the effects of education, urbanisation and modernisation. Increasing urbanisation means declining importance of rural life. Some rural areas are fast becoming rotten and may not be better in the nearest future. Even if they were to reconstitute themselves into the buoyant communities they once were. Western educational influences may not easily permit the village communities to regain their once authoritarian position in ecological conservation. The demise of IK in conservation, therefore, seems final. If this analysis is true, the question that lingers around is why strong consideration should still be given to IK by world bodies like the UNO and the World Bank? The consideration is relevant because such knowledge can be revived. This is evidence from the references made to customary rights, and local population in the country's current forestry law.

It may be argued in favour of indigenous knowledge systems that their procedures form part of the culture of a people and can thus be more sustainably applied within community forest settings than externally motivated procedures. This is the premise on which IK remains important today. Modern methods of conservation have shown very dismal performances and there is the strong belief today in Africa that biodiversity conservation strategies should be contextualised within the *modus vivendi* of people who appropriate the customary rights to forests and other resources. The empowerment of village chiefs through some form of institutional reform that gives them the power to arbitrate some cases of the violation of the local forestry law could be crucial in rekindling interest in IK. The *House of Chiefs* and Customary Courts used to exist as constituent law making and executing bodies in colonial and immediate post-colonial Anglophone Cameroon. These may still be relevant today.

BIBLIOGRAPHY

Bhagwat, S.A. and Rutte, C. (2006). Sacred Groves: potential for biodiversity management. *Frontiers in Ecology and Environment.* 4(10): 519-524.

BUCREP, (2010). Rapport de presentation des resultats definitifs. BUCREP.

Cunningham, W.P. and Saigo, B.W. (2001). *Environmental Science: a global concern.* McGraw Hill, sixth edition.

Draper, M. (2002). *Super African dreams: the mythology of community development in frontier conservation areas in southern Africa.*

Eyong, T.C. (2007). Indigenous knowledge and sustainable development in Africa: case study on central Africa. *In* E.K. Boon and L. Hens, (eds.). *Indigenous Knowledge systems and sustainable development: relevance for Africa.* Kamla-Raj Enterprises.

Githitho, A.N. *The sacred Mijikenda Kaya Forest of Coastal Kenya and Biodiversity Conservation.*

Kassilly, F.N. and Tsingalia, H.M. (2009). Persistence and loss of cultural values of Tiriki sacred groves in Hamisi District, Kenya: Implications for management (RH: Cultural Values of Tiriki Sacred Groves). *Journal of Human Ecology.* Kamla-Raj.

Ministry of the Environment and Forestry, (nd) *A compendium of official instruments on forest and wildlife management in Cameroon.*

Probst, P. and Bühler, B. (1990). Patterns and control on medicine, politics, and social change among the wimbum, Cameroon grassfields. *Anthropos* 85. 447-454.

Sigu, G.O., Omenda, T.O., Ongugo, P.O. and Opiyo, A. *Sacred groves institutions, rule enforcement and impact on forest condition: the case of Ramogi Hill Forest Reserve, Kenya.*

Tangwa Sa'a. *Desertification of the Bamboutos Mountain Forest in West Cameroon Highlands: Advocacy for Action.*

Twarog, S. and Kapoor, P. (eds.) (2004). *Protecting and Promoting Traditional Knowledge: Systems, National Experiences and International Dimensions.* New York and Geneva, United Nations.

World Bank (2004). *Indigenous Knowledge: Local pathways to global development,* The World Bank.

4
Integrating Traditional Knowledge in Climate Change Adaptation

V. Dhulasi Birundha

I. Introduction

Traditional knowledge—the wisdom, knowledge and practices of indigenous people gained over time through experience and orally passed on from generation to generation—has over the years played a significant part in solving problems, including problems related to climate change and variability. Local knowledge is variously referred to as folk knowledge, traditional knowledge, indigenous knowledge, traditional environmental knowledge, indigenous traditional knowledge, indigenous agricultural knowledge, farmers' knowledge, rural people's knowledge, peasants' knowledge, ethno-science, etc. It is based on experience, often tested over centuries of use, and entails many insights, perceptions and intuitions relating to local culture and the environment. It is both dynamic and complex, and is not confined to knowledge about uses and products but also about processes. Local knowledge systems have been found to contribute to sustainability in diverse fields such as biodiversity conservation and maintenance of ecosystem services, tropical ecological and bio cultural restoration, sustainable water management, genetic resource conservation and management of other natural resources. Local knowledge has also been found useful for ecosystem restoration and often has ingredients of adaptive management. Indigenous people that live close to natural resources often observe the activities around them and are the first to identify and adapt to any changes. The appearance of certain birds, mating of certain animals and flowering of certain plants are all important signals of changes in time and seasons that are well understood in traditional knowledge systems. Indigenous people have used biodiversity as a buffer against variation, change and catastrophe; in the face of plague, if one crop fails, another will survive (Salick and Byg, 2007). In coping with risk due to excessive or low rainfall, drought and crop failure, some traditional people grow many different crops and varieties with different susceptibility to drought and floods and supplement these by hunting, fishing and gathering wild food plants. The diversity of crops and food resources is often matched by a similar diversity in location of fields, as a safety measure to ensure that in the face of extreme weather some fields will survive to produce harvestable crops.

Local knowledge is useful for the following reasons (Pandey, 2002a). First, inadequacy of economic incentives to conserve biodiversity as demonstrated recently by Kleijn *et al.* (2001) compel rethinking classical utilitarian approach to resource management. Second, an emerging sustainability science (Kates *et al.*, 2001) will need all stocks of knowledge and institutional innovations to navigate transition towards a sustainable planet. Third, rediscovery of traditional ecological knowledge as adaptive management (Berkes *et al.*, 2000) and need to apply human ecological (Bews, 1935; East, 1936; Muller, 1974) and adaptive strategies for natural resource management (Bates, 2000) offers prospects for scientists to address the problems that beset conservation biologists and restoration ecologists. Fourth, there is an increasing realization that we need innovative ethics and policy to conserve biodiversity and maintain ecosystem functions (Tilman, 2000) and that such ethics need not come from the god; rather, society can cultivate them. Fifth, local knowledge systems are disappearing at a rate that may not allow us even to know what value, if any, such systems had (Cox, 2000; Brodt, 2001; Pandey, 2002a). Finally, in a thought provoking discussion, Cavalcanti (2002) notes that a limitation of economic development is that it is pursued without any considerations—in practice—as to its implications on ecosystems. Village communities and other small-scale societies residing continuously over a territory create, transmit and apply comprehensive knowledge about the resources contained in the territory. In villages where women take active part in natural resource management including agriculture and forestry they develop repositories of local knowledge that is continuously applied, tested and improved over time (Harding, 1998).

II. Climate Change in India

Climate change impacts and associated vulnerability are of particular concern to developing countries, here large parts of the population depend on climate sensitive sectors like agriculture and forestry for livelihood. By adversely affecting freshwater availability and quality, biodiversity and desertification, climate change tends to disproportionately affect the poorest in the society, exacerbating inequities in access to food, water and health. The capacity to adapt is a function of access to wealth, scientific and technical knowledge, information, skills, infrastructure, institutions and equity and therefore, varies among regions and socio-economic groups. Humanity faces exceptional challenge of eroding natural resources and declining ecosystems services due to a multitude of threats created by unprecedented growth and consumerism. Also imperiled is the biodiversity and sustainability of the essential ecological processes and life support systems (Chapin *et al.*, 2000) in human dominated ecosystems across scales (Vitousek *et al.*, 1997). Indeed, human-domination of earth is evident in global change (Schimel *et al.*, 2001; Forest *et al.*, 2002), biodiversity extinctions (Singh, 2002) and disruption of ecosystem functions (Loreau *et al.*, 2001). Ecological problems coupled with unequal access to resources results in human ill-being and threats to the livelihood security of the world's poorest (Balvanera *et al.*, 2001). Climate change, therefore, is intrinsically linked to other environmental issues and to the challenge of sustainable development. In India, climate change will put additional stress on ecological and socio-economic systems that are already facing tremendous pressures due to rapid urbanization, industrialization and economic development. With its huge and growing population, a 7500-km long densely populated and low-lying coastline and an economy that is closely tied to its natural resource base, India is considered to be especially vulnerable to the impacts of climate change. Various studies conducted in the country have shown that the surface air temperature in India is rising at the rate of 0.4°C per hundred years, particularly during the post-monsoon and winter season. Models project that mean winter temperatures will increase by as much as 3.2°C in the 2050s and 4.5°C by 2080s. Summer temperatures will increase by 2.2°C in the 2050s and 3.2°C in the 2080s. Extreme temperatures and heat spells have already become common over Northern India, often causing loss of human life. In 1998 alone, 650 deaths occurred in Orissa due to heat waves. Specific Climate Change Projections for India. An annual mean surface temperature rise by the end of the century, are: ranging from 3 to 5°C under A2 scenario. and 2.5 to 4°C under B2 scenario, with warming more pronounced in the northern parts of India. A 20 per cent rise in all India summer monsoon rainfall and further rise in rainfall is projected over all states except Punjab, Rajasthan and Tamil Nadu, which show a slight decrease. Asia and the Pacific is the world's most natural disaster-prone region with floods, droughts, fires, landslides, cyclones and tsunamis among a range of natural calamities threatening rural lives and livelihoods. Gradual changes in mean temperatures and rainfall are affecting agriculture, forests, marine resources, biodiversity and water availability in the region. Observed changes in climate reported by India Meteorological Department (IMD) for the past 130 years are as follows:

Predictions based on Regional Climate Modelling (RCM) system, known as PRECIS developed by Hadley Centre and applied for India.

Monsoon rainfall at all India level does not show any trend but there are some regional patterns. Areas of increasing trend in monsoon rainfall are found along the west coast, north Andhra Pradesh and north-west India, and those of decreasing trend over east Madhya Pradesh and adjoining areas, north-east India and parts of Gujarat and Kerala (-6 to -8% of normal over 100 years).

A recent study indicates that the intensity and frequency of heavy to very heavy rainfall events is showing an increasing trend during the past 50 years over the region covering parts of Andhra Pradesh, Orissa and Chhattisgarh and Madhya Pradesh.

Mean annual surface air temperatures show a significant warming of about 0.5°C/100 years during the last century and recent data indicate a substantial acceleration of this warming after the 1990s. This is comparable to the global warming trends reported Climate change particularly threatens small-scale rain fed farmers, forest-based and pastoralist livelihood systems, inland and coastal fishing/ aquaculture communities, inhabitants of coastal, floodplain and mountain areas. Reducing greenhouse gas (GHG) emissions from food and agricultural production is a major priority for the region. That climate change will impact Indian agriculture is a foregone conclusion. In what way precisely is not clear, though some clues are emerging through various research projects to capture these signals and assess their impact on crops, livestock, fisheries and other fields. The most widely trusted estimate indicates that crop yields may drop 4.5 to 9.0 per cent as a result of climate change in the next three decades, depending on how much temperatures rise and the impact of extreme climate in different parts of the world. A dent of this level in crop productivity translates into an overall cut of around 1.5 per cent in India's gross domestic product (GDP), given that agriculture accounts for 15 per cent of total GDP.

III. International Recognition of Traditional Knowledge

During the past decade, at least 10 declarations emphasizing indigenous knowledge were made internationally, starting with the Declaration on Environment and Development adopted together with Agenda 21 by the Rio Earth Summit in 1992. In Agenda 21, local knowledge was mentioned as many as 166 times. Since then, over 30 research institutes and archives on indigenous knowledge have been established all over the world. The Albuquerque declaration submitted to COP4 in Argentina is perhaps the first formal declaration emphasizing the need for respecting indigenous cultures and learning from their knowledge of climate change. Indigenous knowledge was also on the agenda of the first (Toronto, 1997) and second (Kuala Lumpur, 2000) conferences devoted to global knowledge for development. The International Indigenous Forums held in Lyon, Hague and Bonn in 1999, 2000 and 2001 respectively further stressed this point. The Scientific and Technical Advisory Panel of the Global Environment Facility, in their workshops in February and April 2002 also underscored the importance of indigenous knowledge and local participation. Because of such initiatives, there is now an increasing appreciation by policy makers of the need to understand local perceptions of climate change problems and solutions. The relevance of traditional knowledge in development of policies and practices that support sustainable management of natural resources has been debated in many international forums. The UN Intergovernmental Panel on Climate Change (IPCC) has recognized indigenous people and local knowledge as key factors towards meeting the challenges of Climate Change. It has been recognized that incorporation of indigenous knowledge into climate change policies contributes to development of effective participatory, sustainable and cost-effective adaptation strategies towards climate change. The Bali climate change conference has strongly recommended the full and effective participation of indigenous people and civil society in all UN processes for climate change. This recommendation has been endorsed by the delegates participated at Indigenous Peoples' Global Summit on Climate Change (April 2009, Anchorage), International Expert Group Meeting on Indigenous Peoples and Climate Change, (April 2008, Darwin), International Public Forum on Indigenous People and Climate Change (April 2008, Darwin), Global Indigenous Peoples Consultation on Reducing Emissions from Deforestation and Degradation (REDD) (November 2008, Baguio City), South-East Asian Indigenous Peoples Regional Consultation on REDD (November 2008, Baguio City), Indigenous Issues and Climate Change Stream at the 11th International Congress of Ethnobiology (June 2008, Cusco), Planning meetings for the Indigenous Peoples Climate Change Assessment Project (June 2008, Cusco and September 2008, Palo Alto) and the first meeting of the Interim Steering Committee (October 2008, Spain). (Ancha Srinivasan, 2004)

IV. Local Knowledge in Climate Change Adaptation

Sustainable development issues are of core concern to the country providing the best mechanism to address most of these issues. Commitment to the UNFCCC as well as many other Multilateral Environmental Agreements highlights India's concerns related to these issues besides other steps that have been taken to address to push socio-economic development and eradicate poverty. The challenges lie in identifying opportunities that would facilitate sustainable development by making use of existing technologies and development policies that make climate-sensitive sectors resilient to climatic variability. This strategy will require developing countries to have access to appropriate technologies, information, and adequate financing. As it is, the coping mechanisms in developing countries are stretched to deal with natural calamities. Adding the climate change dimension implies additional burden and adaptation will require anticipatory planning; failure to prepare systems for projected changes in climate variability and extremes could lead to capital intensive development of infrastructure or technologies that are ill suited to future conditions.

Adaptation to climate change includes all adjustments in behaviour or economic structure that reduce the vulnerability of society to changes in the climate system (Smith, Ragland and Pitts, 1996). The IPCC Third Assessment Report 2001 concluded that current knowledge on adaptation to climate change, and adaptive capacity is limited and emphasized the need for further research on timely, progressive and viable adaptation measures. Further it stressed that adaptation to climate change, sustainable development and enhancement of equity are mutually reinforcing (IPCC, 2001). Most countries in the Asia-Pacific region are heavily dependent on climate-sensitive sectors (agriculture, fisheries, tourism, etc.) and are, therefore, more vulnerable to climate change than other regions. Climate modelling scenarios for Asia indeed predict warmer temperatures, drier soils, rising sea levels, higher frequency of floods and droughts, and consequent shifting patterns of resource availability. For example, floods and erosion have become central components of the life of *char* (unstable alluvial islands) dwellers in Bangladesh, as the islands can disappear into the river within a few days. On an average, a char-dweller aged 45 may have moved eight times because of erosion in Bangladesh (Schmuck-Widmann, 2001). A great deal of traditional and local knowledge already exists in Asia and the Pacific. For example, even in a small area covering 12 villages in Dindigul district of India, as many as 1203 indigenous agricultural practices were recorded (Sundaramari, 2001). Participants at a recently conducted workshop in 2005 on "Community level adaptation to climate change" in Bangladesh also stressed the importance of incorporating indigenous knowledge into adaptation strategies to climate change in developing countries (Hossain, 2005). Indigenous knowledge is not confined to developing countries alone. For example, Ainu tribes in Japan and Maori tribes in New Zealand have reliable traditional knowledge related to their experiences with specific local areas. Policies during the recent past with a greater reliance on modern technologies, however, led to the loss of local knowledge over generations. For example, the widespread cultivation of high yielding varieties of rice made Bangladeshi farmers forget how to select and plant flood resistant local species. The objective is to examine the relevance of local knowledge in facilitating adaptation to climate change, and propose means to integrate local knowledge in adaptation plans.

Indigenous knowledge for adaptation to climate change may be described as knowledge unique to a given culture or society, acquired through accumulation of experiences of local people (farmers, landless labourers, women, rural artisans, herders, etc.) through informal experiments and intimate understanding of the natural systems stressed by climate change and socio-economic development. Both climate change-related stresses such as global warming and socio-economic stresses such as changing lifestyles affect the natural systems. Because local communities have historically and continue to be guardians of these systems, they have developed novel ways of adapting to such stresses. Moreover, local knowledge embodies place-specific experience on climate change, so it allows better appraisal of climate risk factors in production decisions. Although adaptation largely occurs at micro-spatial scales and is site-specific, adaptation plans so far neglected the micro-spatial scale in favour of the regional and national scales. In addition, there has often been too much reliance on top-down prescriptions and not enough understanding of local processes, priorities and the dynamics of local communities. Many experts acknowledge that shortcomings of some adaptation projects were primarily due to ignorance of local knowledge and failure to involve local communities and other stakeholders. Since adaptation to climate change is a cyclical process involving information, planning, implementation and evaluation, indigenous knowledge is crucial in several phases of adaptation policy framework. These include vulnerability assessment, priority setting process, preliminary selection and evaluation of adaptation measures, designing implementation methods for various adaptation strategies, and integration with other adaptation and mitigation strategies. For example, Matamatega of Kaumana (MOK) is local knowledge in Tuvalu for reading of clouds to locate schools of fish in the ocean and to predict approaching disasters (strong winds, droughts, waves) and it can increase community preparedness for disasters associated with climate change (Faavae, personal communication).

V. Integration of Knowledge Systems

Management of natural resources cannot afford to be the subject of just any single body knowledge such as the Western science, but it has to take into consideration the plurality of knowledge systems. There is a more fundamental reason for the integration of knowledge systems. Application of scientific research and local knowledge contributes both to the equity, opportunity, security and empowerment of local communities, as well as to the sustainability of the natural resources. Local knowledge helps in scenario analysis, data collection, management planning, designing of the adaptive strategies to learn and get feedback, and institutional support to put policies into practice. Science, on the other hand, provides new technologies, or helps in improvement to the existing ones. It also provides tools for networking, storing, visualizing, and analyzing information, as well as projecting long-term trends so that efficient solutions to complex problems can be obtained. Local knowledge systems have been found to contribute to sustainability in diverse fields such as biodiversity conservation and maintenance of ecosystem services, tropical ecological and biocultural restoration, sustainable water management, genetic resource conservation and management of other natural resources. Local knowledge has also been found useful for ecosystem restoration and often has ingredients of adaptive management.

Traditional knowledge may indeed complement scientific knowledge by providing practical experience in living within ecosystems and responding to ecosystem change. But, as Berkes *et al.* (2000) note the "language" of traditional ecology is different from the scientific and generally includes "metaphorical imagery and spiritual expression, signifying differences in context, motive, and conceptual underpinnings"). Traditional skills, local techniques and rural craft provide a wide spectrum of knowledge in India, and since "knowledge cannot be fragmented" we have to take the validated local knowledge into account together with science for evolving a robust sustainability science. Sharp boundaries between formal and local systems of knowledge, and natural sciences and social sciences may indeed be imaginary. Scientific community is increasingly realizing that "there is a continuum between artificially dichotomized aspects of science: objective *versus* subjective, value free *versus* value laden, neutral *versus* advocacy".

VI. Indigenous Coping Strategies Towards Climate Change

Indigenous coping strategies of local people resemble scientific methods in their reliance on systematic observation of natural phenomena. Indeed, many indigenous coping strategies which were once considered primitive and misguided are now seen as appropriate and sophisticated (Chambers, 1983). A key component of adaptation to natural disaster and climate change is development of sustainable and alternative farm and non-farm rural livelihood options. This includes ensuring a reliable supply of affordable and environment-friendly bio energy in rural areas to boost farm productivity and non-farm rural enterprises, indirectly helping reduce farming-related GHG emissions. Small-scale cooperative production of bio fuel crops like Jatropha on marginal lands in India and Thailand has a large rural job-creation potential. Arm family members are preparing themselves to cope with natural disasters and climate change. This includes promoting public awareness, building cyclone shelters, providing immediate relief and rehabilitation, alternative cropping, storing food and essential supplies, and developing partnership with other sectors.

Indigenous coping strategies range from simple methods such as maintaining hardy wild foods (e.g., consumption of wild fruits and vegetables that survive during droughts) to more complex methods of reducing risk through cultivation of more than one type of grain staple, mixed land use, intercropping and mixed cropping. Adaptive capacity of such traditional mixed cropping and agro-forestry systems has been shown to be much higher than that of modern farming systems. Using local germplasm highly acclimated to withstand harsh climates is one innovative strategy. Farmers in India base their cropping pattern decisions on Krishi Panchang (agricultural Vedic calendar), which includes a climate forecast for an entire year based on planetary positions. Bio-dynamic agriculture, a system initiated by an Austrian philosopher, Rudolf Stiener, and practiced now in nearly 25 countries also depends on similar calendars. For example, farmers in Gujarat, India decide to cultivate groundnut and cotton if there is an early onset of rains, and pulses and millet if there is delayed onset. The scientific basis of some of these practices remain unclear however. For example, only recently Orlove *et al.* (2000) found out the basis of indigenous knowledge of potato farmers in the Andes, who usually forecast inter-annual variations in summer rainfall and in autumn harvests based on changes in the apparent brightness of stars in Pleiades around the time of the southern winter solstice. Farmers minimize the risk from low rainfall during the growing season in an El Nino year by varying the planting date.

Indigenous Coping Strategies in the India

Agricultural Sector

Instead of relying on high input agriculture by using hybrid seeds, chemical fertilizers and pesticides, some farmers still trust on the force of **traditional soil and crop management practices** in the villages of **Rajasthan**. Such practices consider the highly variable environment and the fragile ecosystem in a far better way than agricultural activity, solely aiming at maximizing outputs. Such sustainable practices include among others. Some innovative farmers are partially practicing **organic farming** on their plots by using compost based fertilizer and bio pesticides, produced out of locally available weed. Such practices help maintaining an intact biodiversity, which provides resilience to the ecosystem against adverse impacts of climate change. Soil fertility and moisture content of the fields are improving rapidly with reduced input of chemical fertilizers. Farmers are practicing **summer ploughing** one of the important traditional methods to control insects and diseases and helps to kill weeds, hibernating insects and disease causing organisms by exposing them to the summer heat. Many farmers are still using **indigenous seeds** which are more resistant to drought as compared to hybrid ones and do not need the same amount of fertilizer input to produce decent yields. In addition, farmers can save considerable amounts of money by conserving their own seeds, which buffers financial losses in case of crop failure. To offset crop failure arising from rainfall variability and unpredictability, farmers cultivate several hardier or **drought-tolerant** types of the same crop species. Many prefer the use of local grains such as millets and sorghums that are more drought-resistant than many maize or wheat varieties. In rain deficient years, such crops are considerably better in term of yields, than the more water demanding species. Due to reduced water availability, some farmers have stopped growing crops, which require greater amount of water such as cotton and wheat. They have successfully **shifted cropping patterns to species requiring less water**. They started growing cumin seeds, gram, black gram, and some oilseed crops like mustard. A diversity of crop types and varieties are grown in rotation and in different areas of farm properties, which spreads the risk of losing an entire year's production. The important crops grown in the **mixed cropping system** are Pearl millet, Sesame, Green gram, Sorghum and Maize. By necessity, farmers shift their sowing periods according to the weather. Earlier, farmers were taking decisions about their cropping pattern and sowing time on **Krishi Panchang** (agricultural Vedic calendar), which includes a climate forecast for an entire year based on planetary positions. Such traditional weather forecast is not working appropriately anymore and farmers adapted to changing weather conditions by taking more spontaneous decisions. In the absence of efficient weather forecast such decisions are, however, risky and many farmers reported to be helpless facing increased weather variability.

But in **Andhra Pradesh** farmers are **adjusting their cropping calendar to changing weather conditions.** According to them, the beginning of the monsoon season is constantly shifting backward, reducing the length of growth period for crops. Traditionally, end of April used to be the time for sowing crops. Farmers could expect first monsoon rains soon after and rainfall was mostly constant until August. Now-a-days, the rains are frequently delayed by several weeks and are less regularly distributed over the rainy season. Accordingly, sowing time has gradually shifted to mid-May or even beginning of June. This is adaptation per necessity, as a reduced growing season means less crop yields. Farmers do not have a choice as sowing early without irrigation would produce crop failure. As the growing seasons get shorter, farmers are constraint to adapt their cropping patterns. Where no irrigation is available, farmers have stopped growing paddy as the crop failed too often because of deficient rainfall. Pearl- and Finger-Millet as well as **short duration pulses (red gram) are planted** instead. These crops are better adapted to local conditions: They are more heat resistant, need less water and have shorter growing cycles. This allows for increased food security for the farmers and their families and preserves scarce water resources. **Mixed cropping** is a promising adaptation strategy for small-scale farmers to reduce risk of complete crop failure. Some farmers have started growing groundnut together with pigeon pea, which mitigates the risk from reduced groundnut yields. Both crops are well adapted to water scarce conditions prevailing in Anantapur and grow well even on relatively poor soils. Mixed cropping of different vegetables like eggplant, tomatoes and ladyfingers together with corn and millets is working well too. Similarly, the farmers of **Uttarakhand** are also adapting. diverse farming and mixed cropping The advantage is that with climate variability and uncertainty, due to the combination of different species and varieties, it is unlikely that all crops will be negatively affected and the overall productivity will not be zero even when there is drought or excessive rainfall. The advantage of mixed cropping is that it will ensure some income even if one of the crops fails due to pests, diseases, drought or any other natural calamity. Today, some 126 varieties of rice, 8 of wheat, 40 of finger millet, 6 of barnyard millet, 110 of kidney beans, 7 of horse gram, 8 of traditional soybean, and 10 of French beans, are being grown with no chemical inputs. The characteristics of each (e.g. growth, resistance, special properties, etc.) are being carefully observed. Varieties with desirable properties, like high productivity and resistance, are being propagated amongst the farmers (Jardhari and Kothari, 1997). For farmers, mixed cropping ensures food security as well as income. Fox example, when farmers grow only rice they are more vulnerable because if the rice crop fails due to, for example, drought, they are left with nothing. From the farmer's perspective this system is more or less free of pests and diseases. Even if it exists, only one or two crops in the mixture are affected. The rich biodiversity protects the other crops. Even in the case of heavy wind or storm, only one or two crops are affected. Through the cultivation of drought-resistant crops farmers are adapting cropping patterns to less water demanding species as most important actions to environmental changes. Farmers generally grow 7-8 varieties of rice. Of these, 2-3 varieties are for non-irrigated fields. The farmers explained that they grow more than one variety of rice because different varieties are suitable for different types of fields. Some of the traditional rice varieties require relatively little water compared to new high yielding varieties, it is being in certain cases the traditional varieties are being promoted by the government. The ***"Baranaaja"***, as the cropping system is called, can be described as a **'simultaneous system'** in which components are grown in mixtures and the interactions are both spatial and temporal. It uses cropping patterns involving 12 or more food crops grown in 'synergetic' combinations such as *mandua* (finger millets), *ramdana* (amaranth), *rajma* (common beans), *ogal* (buckwheat), *urad* (green gram), *moong* (black gram), *naurangi* (mix of pulses), *gahath* (horsegram), *bhat* (soybean), lobiya (French beans), *kheera* (cucumber) and *bhang* (cannabis) are grown together in a way that optimizes productivity and maintains soil fertility. The "synergic" combination of many different species does not deplete the fertility of the soil and the land does not have to be kept fallow. This agro biodiversity is one of the important aspects of the sustainable food system in the region. This system of bio-farming enables the farmer to get some benefit from certain varieties even in case of damage to other crops. Another interesting example from the farmers was a very old method of transplantation of rice in the fields. This method is known as **"Thakuli" (plate) method**, which is similar to the SRI method (System of Rice intensification). In this method, younger and fewer seedlings are transplanted with wider spacing to promote greater root growth and higher yields. The farming system is highly integrated, e.g. the soil fertility is maintained *via* livestock which also provide the energy for the different agricultural operations. Continuous flooding is not required and the fields are fertilized with farmyard manure. This practice reduces water use and seed requirements (less seeds are sown). Farmers usually apply Farm Yard Manures (FYM) which involves use of fully decomposed organic matter of livestock dung, straw, grasses, left over feed etc. The leaves used for animal bedding are also used for making organic manure. FYM is prepared within a period of about 3-4 months of continuous open-air decomposition of cow dung with other leaf litter.

Farmers also try to plant nitrogen fixing or fruit bearing trees in strategic location in order to limit erosion of their fields and increase soil fertility.

Farmers in **Orrisa** cultivate several **varieties of short duration and long duration upland paddy, especially suited to grow on the middle region slopes.** They also grow some of the most exquisite varieties of scented rice, the most famous amongst them being 'Kala Jeera' so called because the paddy is black in colour before being de-husked. Apart from this, they have short and long duration varieties of Ragi and the less common millets, including fox tail millet, pearl millet, sorghum, and others. The same is true for pulses: several varieties of broad bean, arhar, cow pea, rice bean, urad, and a local variety commonly called 'Baeil' are grown at different altitudes of the hill slopes.

Maharastra farmers increasingly recognize that the existing farming system with its excessive dependence on soybean and cotton is degrading the natural resource base. In addition, these crops are not very tolerant towards water stress and without irrigation fail too often. In order to overcome the overdependence on these two crops, some farmers are planting traditional dry land crops again (arhar, sorgum, green gram, black gram, millet and pigeon pea). As adaptation to climatic variation: an alternative farming model known as"**Self Reliance Farming"** in Wardha, is based on low external inputs and high self-reliance of food and farming inputs (e.g. seeds, fertilizers, pesticides and fungicides, equipment and machinery). The model was developed for land units of 1-2 hectares under rain-fed conditions, which previously cultivated cotton. It aims at providing a resilient agro-ecosystem with inbuilt resistance against seasonal weather variations by careful selection of different crop varieties. As the total growing period in the area is only 6 months, crops which are able to grow side by side are necessary. The model uses a combination of cash crops and food crops to diversify the farmers' food sources and achieve nutritional security, as well as economic benefits through sales of cash crops. Using this practice has several advantages for farmers: There are less pest-attacks and diseases, so the need for chemical pesticides is reduced. Additionally, by inter-cropping, the plants cover the soilsurface better, so are able to conserve more moisture in the soil and to convert more sunlight into food and fibre. Due to changes in rainfall patterns, many farmers try to minimizing losses from crop failure through increased use of drought-tolerant crops, including pigeon pea, sorghum, green gram and black gram.

Water

Indigenous systems of rain water harvesting such as **surangas** (man-made caves for water) in Kasargod district of Kerala, **madaka**s (traditional percolation ponds) in coastal Karnataka, and **johads** in Rajasthan in India are unique ways to cope with droughts. Local farmers also devised ingenious ways of dealing with climate extremes. Farmers in South India, for example, manipulate microclimate in betel vine gardens to produce high quality broad, light green, feathery leaves. Farmers of Nubra valley in the western Himalayas grow grapes under severe cold desert conditions by creating a warmer microclimate of the basin. This is achieved by filling pits with locally available crushed stone, bricks, grass, hay and soil, and by covering the vines with warm clothes, gunny bags or wooden baskets. Char dwellers in Bangladesh build mounds and construct their dwellings to minimize damage from floods and plant catkin reed (*Saccharum spontaneum*) to protect chars from erosion. Catkin reed, which grows with the water level and survives floods, promotes stabilization of new land through fixing the fine, fertile sediments. In addition, it serves as the most important fodder during the monsoon season, fuel and building material (Schmuck-Widmann, 2001). Warabandi system of water allocation is a well-recognized practice to cope with droughts in many parts of India, Pakistan and Nepal. Water temples in Bali, Indonesia, are now widely recognized as indigenous strategies that contribute to a delicately balanced system of cooperation between neighbouring farmers. In **Rajasthan** Communal water storage facilities, called **bawari, jhalara and nadi**. This method is centuries old but still is a cost-effective way to get water for irrigation and livestock without overusing the groundwater. When properly maintained, such structures mitigate drought impacts at least partially. In **Andhra Pradesh:** No more wells have been dug in the village since 2004 and overuse of groundwater is reduced to a sustainable level, but paddy crops were replaced with irrigated dry crops such as groundnut, pigeon pea, pulses and horticulture crops. Investment in drip irrigation or sprinklers is a good way to reduce water consumption in agriculture. However, most of the time farmers are not capable of making investments of that scale. Farmers were not aware of all government schemes, including schemes that had provisions of subsidies for such irrigation facilities. To make efficient use of available rainwater, farmers practice mulching by applying crop residues on their fields in order to keep soil temperatures relatively low and preserve available soil moisture. Trenches are built around the field to collect and divert as much rainwater as possible to the arable land. In steeper parts such trenches are built horizontally to the slope in order to limit erosive forces of rainwater and augment water uptake of the soil. Canals, leading from slopes towards the fields, are dug in order to maximize utilisation of rainwater and limit run-off. In **Maharashtra** community-based water management initiatives can have significant impacts on water availability (Poonam Pande, 2011).

VII. Incorporating Traditional Knowledge in Practice

Any attempt, endeavouring to integrate traditional knowledge for biodiversity conservation and sustainability of natural resources should be based on the principle that traditional knowledge often cannot be dissociated from its cultural and institutional setting. Regarding the cultural and institutional the following suggestions may be useful:

1. Each programme aiming at the promotion of traditional knowledge should be based on the recognition that natural resource rights and tenurial security of local communities forms the fundamental basis of respecting traditional knowledge.
2. More attention is needed on protection of intellectual property rights of traditional people.
3. Innovative projects may need to be developed that aim at the enhancement of the capacity of local communities to use, express and develop their traditional knowledge on the basis of their own cultural and institutional norms.
4. Traditional knowledge and traditions can contribute to the preparation of village microplans, which are prepared for eco-development, joint forest management and rural development. The plans should be based on both geographic and traditional community boundaries rather than only on administrative boundaries.
5. There is a clear need to integrate traditional and formal sciences for participatory monitoring, and taking feedback to achieve adaptive strategies for management of natural resources.
6. The goal of climate change adaptation policies should be to enhance adaptive capacity and facilitate actions that reduce vulnerability and improve resilience of local people to climate change, which in turn can advance sustainable development and equity.

   One of the promising methods to design socially benign and culturally appropriate adaptation policies to document and understand local strategies to cope with climate change. This is in part because indigenous societies accumulated valuable knowledge that helps in understanding human-environment relationships for optimum utilization of resources and to regulate human environmental impacts.
7. Encouraging the documentation of indigenous knowledge and its use in natural resource management. Such documentation should be carried out in participation with the communities that hold the knowledge. Due attention should be given to document the emic perspectives regarding IK rather than only the perspectives of professional outsiders. The documentation should not only consist of descriptions

of knowledge systems and its use, but also information on the threats to its survival.

8. Developing clear and concise educational material on traditional knowledge systems to be used in communication programmes to impart information regarding the merits and threats to indigenous knowledge systems to both policy makers and the general public.
9. Developing curricula and methods for providing formal training and education in traditional knowledge systems to agencies, researchers and practitioners who work in collaboration with communities.
10. Developing research projects aimed at assessing the possibilities and constraints of using traditional knowledge under specific conditions. Such research projects should move beyond the first generation research projects, which aimed at demonstrating the value of local knowledge systems by focusing on successful cases of application. Second generation research projects shall focus on comparing application of knowledge systems across a range of circumstances and across disciplines to craft the traditional sustainability science. Developing new methods for incorporating local knowledge systems in natural resource management regimes through action research.

VIII. Conclusion

Indigenous knowledge is a powerful asset and social capital in many countries of Asia and the Pacific, which are vulnerable to climate change. Policies incorporating such knowledge are more likely to be acceptable and successful in the long run. Both indigenous knowledge and climate change adaptation are site-specific. Moreover, indigenous knowledge is dynamic in the sense that people continue to generate new local knowledge on various coping strategies.

Indigenous knowledge and climate change adaptation are multi-level (individual, community, local, regional, national and international) and multi-sector endeavours. The adaptation options and strategies that we design must be, therefore, both proactive and reactive, relying on local experimentation and innovation, and community involvement. Indigenous communication channels increase the rate of dissemination and utilization of adaptation strategies, as they provide mechanisms for facilitating understanding and communication. Indigenous coping measures tightly link local knowledge, social responsibility and equity. A shift in paradigm from "top-down approach" to "bottom-up participatory approach" is, therefore, vital to improve adaptive capacity of people and for ensuring effective technology transfer. Active participation by the local people who are most affected, and who accumulated knowledge on how to survive climate extremes is important to develop any practical and relevant climate change adaptation plans. The partnership approach must involve local people at all stages of the technology development and policy formulation process, including the generation, dissemination and utilization of technology, as well as research planning, implementation and evaluation. Integration of local knowledge in adaptation strategies should be a value-added approach and must benefit everyone involved thereby leading to win-win situations. It is worth remembering, however, that climate change is a global phenomenon while adaptation is largely site-specific. As site-specific issues require site-specific knowledge, indigenous bottom-up strategies are likely to be more successful than top-down approaches in adapting to and cope with climate change. Indeed successful adaptation activities are often built upon local knowledge. Despite such importance of local knowledge in enhancing adaptation, it has been largely ignored by policy makers due to inefficiencies arising from institutional weaknesses, over-emphasis of the public-sector, centralized decision-making and neglect of the central role of producer organizations, local communities and the importance of family farms over time. Indigenous knowledge systems were also considered to be inferior and denied role in designing policies. Inadequate consideration of different local practices was often blamed for poor success in introducing "scientific adaptations". Optimal integration of local knowledge into climate change adaptation plans might also result in more culturally appropriate options, and present a more holistic and integrated perspective. Technologies and tools such as GIS must be taken advantage of in integrating indigenous knowledge in adaptation plans.

REFERENCES

Ancha Srinivasan (2004). *Local Knowledge for Facilitating Adaptation to Climate Change in Asia and the Pacific: Policy Implications*, IGES-CP Working Paper (2004-002) November 2004.

Bates, D.G. (2000). *Human Adaptive Strategies: Ecology, Culture, and Politics*. Allyn and Bacon, pp. 238. 2nd ed.

Berkes, F., Colding, J., and Folke, C. (2000). Rediscovery of traditional ecological knowledge as adaptive management. *Ecological Applications* 10: 1251-1262.

Brodt, S.B. (2001). A systems perspective on the conservation and erosion of indigenous agricultural knowledge in central India. *Human Ecology* 29: 99-120.

Cavalcanti, C. (2002). Economic thinking, traditional ecological knowledge and ethnoeconomics. *Current Sociology* 50: 39-55.

Cox, P.A. (2000). Will tribal knowledge survive the millennium? *Science* 287: 44-45.

Chambers, R. (1983). *Rural development: Putting the Last First*. Longman Scientific and Technical, co-published with John Wiley and Sons Inc. New York.

Harding, S. (1998). Women, science, and society. *Science* 281: 1599-1600.

Hossain, M. (2005). *Local tricks to tackle climate shifts*. SciDev.Net 4 Feb. 2005.

IPCC (Intergovernmental Panel on Climate Change) (2001). *Climate change 2001: Impacts, adaptation, and vulnerability. intergovernmental panel on climate change.* Cambridge, UK: Cambridge University Press.

IPCC (Intergovernmental Panel on Climate Change) (2007). *Climate change 2007.* Contribution of Working Group I to the Fourth Assessment Report of the Intergovernmental Panel on Climate Change. Cambridge, UK and New York, USA, Cambridge University Press.

Kleijn, D., Berendse, F., Smit, R., and Gillissen, N. (2001). Agri-environment schemes do not effectively protect biodiversity in Dutch agricultural landscapes. *Nature* 413: 723-725.

Kates, R.W. *et al.* (2001). Sustainability Science. *Science* 292: 641-642.

Muller, C.H. (1974). Human ecology. *Science* 183: 368.

Orlove, B.S., Chiang, J.C.H., and Cane, M.A. (2000). Forecasting Andean rainfall and crop yield from the influence of El Nino on Pleiades visibility. *Nature* 403:68-71.

Poonam Pande and Kaspar Akermann (2011). *Adaptation of small scale farmers to climatic risks in India, Sustainet India,* www.sustainet.org viewed on May 23, 2011

Salick, B. and Byg, A., (eds.) (2007). *Indigenous peoples and climate change*. Oxford, UK, Tyndall Centre for Climate Change Research.

Schmuck-Widmann, H. (2001). *Facing the Jamuna River: Indigenous and engineering knowledge in Bangladesh.* Bangladesh Resource Centre for Indigenous Knowledge, Dhaka. 242 pp.

Smith, J.B., Ragland, S.E. and Pitts, G.J. (1996.) A process for evaluating anticipatory adaptation measures for climate change. *Water, Air and Soil Pollution*, 92: 229-238.

Sundaramari, M. (2001). *Adoption and perceived effectiveness of indigenous agricultural practices in different farming systems*. Ph.D. Thesis, Gandhi Gram Rural Institute, Tamil Nadu, India.

Tilman, D. (2000). Causes, consequences and ethics of biodiversity, *Nature* 405: 208-211.

5

Reclaiming Wastelands for Agro-Afforestation in Amethi

Hridai R. Yadav

1. Statement of the Problems

Excessive population growth in India and the growing demand for fuel wood, fodder, fibre, fruits, fisheries and foodgrains (F^6) has led towards eco-socio-economic imbalances over exploitation of natural resources. It has also caused environmental degradation and increase in the area under wastelands. The land-sat-imagery data fairly well documented report shook the nation to know that we have lost more than 50 per cent of our forest cover. The then Prime Minister Rajiv Gandhi in his address on January 5, 1986 mentioned the establishment of "National Wastelands Development Board" and Prime Minister talked about ecological and socio-economic crises but what needs to be underlined is that the crises affects to the poor much ore significantly. The major impact of deforestation and ecological degradation that has taken place has been largely on the rural poor because half of our nation depends for its fuel wood and fodder on the common lands and it is the degradation of these common lands which has made not only their poverty much more acute but the quality of life of the poor. It has been recorded that out of 75 million hectares which are considered to be forest cover area, about 40 million hectares are considered without sufficient forest tree cover. But it is not only the forest area but we have lost 26 per cent to 52 per cent of our grazing lands so here again the impact has been largely on the poor. Thus, on the one side, the productive forest resources and the grazing lands have reduced by half and on the other the population has increased very significantly including cattle population. Thus it is evident that the reduction in the land resources, its productivity, its degradation and the land is not producing the fuel wood and fodder which are being required by the people. The tremendous growth in cattle and human population has made the crisis very significant. Thus the reclamation of wastelands and restoration of degraded lands have alluded and we have to bring our country on ecological balance. To achieve the ecological balance we need at least 33 per cent of our land resources under forest cover, at the moment we have less than 11 per cent. It is absolutely necessary and imperative that we do afforestation by 50 million hectares every year. It was not in terms of the feasibility of increasing from four million hectares which was done in the Sixth Plan to 25 million hectares in Seventh Plan.

The target for afforestation on wastelands, many people said, was unrealistic, on the other hand, in one of the meeting the then Prime Minister said "can we afford to do it, do we have the resources do it, can we afford not to do it?" Really that was the key question and that was in the mind of Late Rajiv Gandhi. Massive afforestation could have been achieved through peoples participation and peoples movement. The Agro-Horti-Afforestation programme was not implemented as per the requirement of the growing population, due to which the deforestation crisis is so disparate, that the crisis is so real that we must involve everyone in this movement.

The over-exploitation of natural resources for meeting the increasing demand of growing cattle and human population for fuel wood, fodder, fibre, fruits, fisheries and foodgrains (F^6) and other items has led to environmental degradation and rapid expansion of wastelands.

There are more than 175 million hectares degraded land out of the total area of only 328 million hectares. It has been observed that our country is losing 1.3 million hectares forest per year. The continued deforestation has brought us face to face with a major ecological and socio-economic crisis. It must be halted.

The present fire wood requirement of our country is more than 130 million tonnes per year of which only 39 million tonnes are fetched from the forest. The life of tribals and scheduled caste is interwoven with the forest and unless there is replenishment through people afforestation there will be complete breakdown of eco-socio-economic system.

Magnitude of the problem of wastelands and gravity of the situation are evident from the fallowing points:

- In forestry sector out of 75 million hectare land 40 million hectare land is degraded, 30 million hectare is without forest tree cover and million hectare is with only shrubs.
- In agriculture sector out of 143 million hectare area 40-60 million hectare land is considered to be degraded land.
- The pasture and grazing lands and other public lands are the most degraded ones because they have also been the most neglected areas.
- Our country is loosing about 6000 million tonnes of soil requirement to 5.37 to 8.4 million tonnes nutrients per annum which represents a production loss of 30 to 40 million tonnes of food grains per year.
- Approximately 40 million hectares area is flood prone and 260 million hectares area is drought prone, besides soil erosion by wind and water while there are several outer causes of the soil degradation.

Presently more than 1000 million population of our country requires 225 million tonnes of food grains, 30 million tonnes of sugar, 965 millions kg tea, 159 million kg. coffee, 10.2 million tonnes vegetable oils, 17 million bales cotton and 70.0 million cubic metre wood. To achieve this target 40 million hectares area should have to be brought under food grain production, for fuel and other needs and 10 million hectares each for the production of crops and fodder along with 6 million hectares land for increased urbanization. The estimate fuel requirement of our country is 230 million cubic metre.

It is evident from the above observations that points to the urgency of the following:

- Strengthening and launching of educational programmes on wastelands development and agro-afforestation by and for the beneficiaries, government officials, as well as the general public, so that the minimum requirement of deprived rural poor people, scheduled caste, scheduled tribes are met;
- A massive campaign for increasing the area under Agri-Horti-Afforestation for fuel wood, fodder, fibre, fruits, fisheries and foodgrains (F^6), vegetables etc.;

- Development of appropriate suitable technologies for different climatic soils, topographic conditions for utilization of wastelands; and
- Identification of optimum patches of wastelands in different locations with respect to wastes and soil resources for maximization of benefits to rural poor.

The economic progress should not be achieved on account of exploiting **developmental resources of tomorrow**. If the progress till date is evaluated then it can be easily concluded that the over all progress have been obtained through exploiting available natural resources.

The sustainable development have been defined as, "when all strategies for development have basic theme of environmental sustainability ecological balance, food, fodder, fuel wood, security for employment generation raising income level and removing regional disparities is know as sustainable development."

In view of the hereditary rights our generation is authorised to **utilize the left out natural resources** from our ancestors but not snatching the happiness sources of our future generation. Sustainable or inclusive development should be viewed on the basis of intergenerational system. Thus the inclusive development is cattle to fulfil the human needs for future generation and maintaining the eco-socio-economic imbalances of the region.

2. Objectives of the Study

The major important objectives of the study are as follows:

- To analyze the distribution of wastelands at village level in Amethi.
- To describe the factors responsible for the growth of wastelands.
- To explain the important parameters for the reclamation of wastelands.
- To analyze the economic of reclaiming wastelands.
- To describe the policy planning, strategies for the reclamation of wastelands.

3. Distribution of Wastelands in Amethi

The challenging problems of wastelands and its reclamation should be taken as a challenge to meet the increasing demand of growing cattle and human population in the form of fuel wood, fodder, fibre, fruits, food grains, fisheries and vegetables (VF[6]), with the view to restòre ecological and environmental degradation and to improve the socio-economic crisis of the area.

The wastelands are those lands "which are uncultivable or is presently lying unutilized due to different constraints but it had been used previously, which is giving very actual return, i.e. it has low economic potential and it is ecologically unstable or whose top soil has completely lost its fertility status, which has developed toxicity and is, therefore, unfit for the growth of crops and trees due to environmental or anthropogenic problem, has been advanced and no further use has been found for it."

The classification of wastelands have been derived from land use categories taking cultivable waste and fallow land.

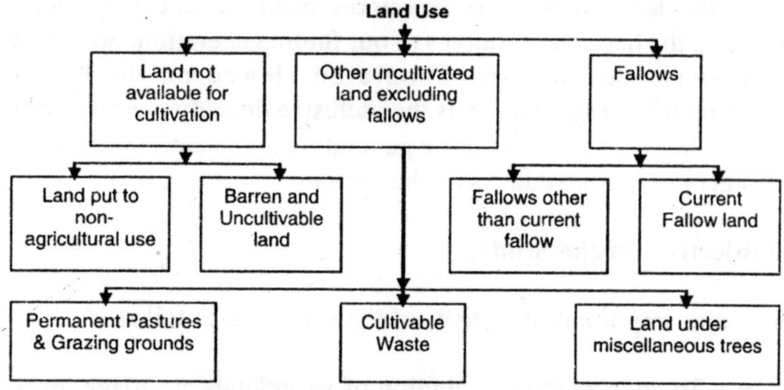

The cultivable waste have been subdivided into waterlogged, usar, banjar, kankarili, ravine and others types of wastelands while fallow land excluding fallows other than current fallows has been subdivided into old fallow land and fallow land excluding current fallow land.

Thus, eight categories of wastelands are available in Sultanpur district, while only six categories of wastelands, viz. waterlogged, usar, banjar, old fallow, fallow and other types of wastelands are found in Amethi Block of the District of Uttar Pradesh.

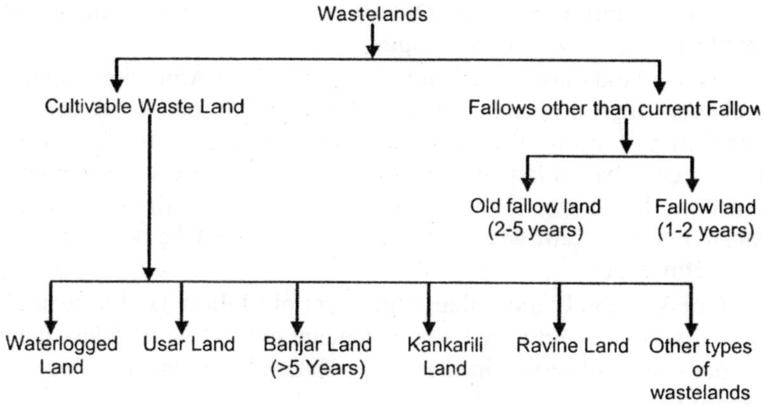

The wastelands are significantly distributed in Sultanpur District. There are 103,608 hectares of wastelands in Sultanpur District. The fallow lands are found comparatively very high covering 41,044 hectares land and usar land 14,601 hectares. The waterlogged land is found in 11,291 hectares area and banjar land in 9,594 hectares area. Similarly, significant area under ravine land along with the Gomati river is found covering 5,510 hectares land and the miscellaneous other types of wastelands are found is 1,925 hectares while very low area covering 748 hectares kankarili land is observed in Sultanpur district.

Proportionately the wasteland is very significantly distributed in Amethi Block at Village level. There are 11.36 per cent total wastelands found in Amethi Block. Comparatively fallow land area is very high which is 3.87 per cent of the total area of the Block. The old fallow land is found in 2.56 per cent area while usar land is observed in 1.89 per cent area. The other types of wastelands in Amethi Block are recorded 0.35 per cent and waterlogged land covering 0.47 per cent area in Amethi Block of District Sultanpur.

At village level water logged land is found very high in Chaturbhujpur, Dehra and Ramdaipur villages while very low water logged lands are observed in Kherauna, Passanwa, Kushi Tali, villages of Amethi Block.

The usar lands are comparatively very high more than 3 per cent in Bhushari and Umpur Gana Patti, Trilokpur, and Sarai Khema villages while very low percentage less than 1.0 per cent usar land is observed in Kherauna, Parsawa, Hathkila, Loniapur, Katara Maharani,

Kushitali, Naraini, Saraiya, Duban and Himmatgarh villages of Amethi Block of District Sultanpur.

Banjar lands are significantly distributed in Amethi at village level. Very high percentage of banjar land more than 4 per cent is found in Loniapur, Raipur Phulwari, villages while very low percentage of banjar land less than two per cent is observed in Gaderi, Bhushari, Ramgarh, Himmatgarh, Agahar, Maharajpur, Saraiya, Daban, Loharta, Katra Maharani, Sarai Khema, and Kherauna villages of Amethi Block.

Proportionately more than 5 per cent old fallow land is seen in Hathkilla village and very low per centage old fallow land less than one per cent is observed in Kherauna, Katara Phulkunwar village of Amethi Block.

Comparatively very high percentage more than 5 per cent of fallow land is seen in Hathkila and Sarai Khema, villages, less than 1.5 per cent fallow land is observed in Parsawan, Benipur, Ramdaipur, Trilokpur, Loharta, Mochwa, Kekwa, Mahmadpur villages of Amethi Block of Sultanpur in District.

Other types of wasteland are found very high more than 5 per cent is Kherauna, Parsawan, Benipur, Katara Phulkunwar villages, while very low less than two per cent other types of wastelands are seen in Sarai Khema, Trilokpur, Tala, Loharta, Saraiya Duban and Himmatgarh villages of Amethi Block.

The total wastelands are very significantly distributed in Amethi Block. Very high percentage of wastelands more than 15 per cent is seen in Dedhpasar village followed by Hathkila (14.32%), Loniapur (14.75%) villages while very low percentage of wastelands less than 8 per cent are recorded in Kushi Tali, Loharata, Korari Girdharshah, Nuwawan, Kakwa, Saidpur, Purabagan, Ramgarh, Himmatgarh and Goderi villages of Amethi Block of Sultanpur District.

Thus it is observed that the wastelands are significantly distributed at village level not only in Amethi Block but Sutanpur district as a whole, which has become a challenging problem to the planners and developmental agencies engaged at Sultanpur.

It has been recorded during the field survey and action fieldwork programme that the wastelands of Sultanpur Amethi Block has been left out of cultivation for very smaller reasons and ignorance of the cultivators. Such wastelands can be easily reclaimed adopting latest scientific and technical methods and applying C-2 scientific technology, i.e. the scientific system available within the village or household itself. The adoption of scientific cropping system/pattern, proper ploughing techniques appropriate doses of organic and inorganic amendments in the form of fertilizer and application of insecticides and pesticides appropriate high yielding variety of seeds, plants, proper management system including irrigation and post harvest technology will be helpful in bringing out the wastelands under proper use to grow fuel wood, fodder, fibre, fruits, fisheries and food grains to meet the increasing demand of growing cattle and human population on the one hand and on the other by restoring the ecological and environmental degradation and imbalance of the area and improving socio-economic crisis of the peoples of Amethi.

4. Factors in Wastelands

The relationship between man and nature is symbiotic. The combination of human and natural factors play an important role in the spatial distribution of wastelands. The correlation matrix and stepwise regression analysis statistically proves that slope and drainage density are the dominant variables in explaining the variations in waterlogged land at village level in Amethi. The main important variables in explaining the proportionate distribution of usar land is the *p*H value of the soil and scheduled caste population followed by land concentration. The important variable in explaining the distribution of banjar land are the *p*H value of the soil, followed by scheduled caste population. Land concentration are the major dominant variables in explaining the variations of old fallow land at village level in Amethi Block. The *p*H value of the soil and drainage density are the main dominant variables in explaining the proportionate distribution of fallow land at village level in Amethi. The land concentration is one of the main dominant variable followed by the scheduled caste population and *p*H value of the soil in explaining the distribution of other types of wastelands.

The main dominant variables in explaining the proportionate distribution of wastelands at village level in Amethi Block are the *p*H value of the soil, land concentration, scheduled caste population and behaviour of water table. Thus the variables, which feature dominating in explaining the variations in wastelands, are the natural factors which are *p*H value of the soil, drainage density, behaviour of the water table and human factors which comprises mainly the land concentration and scheduled caste population. Thus, lastly it can be said that natural and human factors including some of the other variables directly or indirectly responsible for the proportionate distribution of wastelands at village level in Amethi Block, District Sultanpur.

5. Reclamation of Banjar Land

The reclamation capability classification of wastelands had been made based on the fertility status of the soil, and limitations, their capability of producing agricultural crops horticultural and afforestation plantations. It has been observed that the fallow land (1-2 years) are very easily reclaimable for Agri-Horti-Afforestation programme. The old fallow land (2-5) years are easily reclaimable for Agriculture, Horticulture and Forestry plantation. The banjar land which has been left out of cultivation for more than 5 years are reclaimable with certain difficulty for forestry-horticulture-agriculture cropping. The usar land is reclaimable with difficulty for forestry and horticulture plantations. The waterlogged land is reclaimable with moderate difficulty for fisheries, forestry and horticulture plantations.

The eco-sustainability through bio-scientific reclamation of various types of wastelands for Agri-Horti-Afforestation programme for employment and increased generation has been practically demonstrated in the village Benipur of Amethi Block. The cropping quality of land at plot level including land quality based on Khasra map has been depicted in the map. The Agri-Horti-Afforestation programme is technically and economically viable, scientifically, suitable, socially-acceptable because it will fulfil the increasing demand of growing cattle and human population.

The reclamation of banjar land has been demonstrated in the village Benipur where 3.27 per cent banjar land has been found. The reclamation of Banjar land in the village Benipur has been practically demonstrated in one acre banjar land with the continued involvement of the beneficiary. The peoples participation in the reclamation of banjar land has been ensured:

- The reclamation of banjar land has been initiated through levelling, binding, top dressing of fertilizer, irrigation.

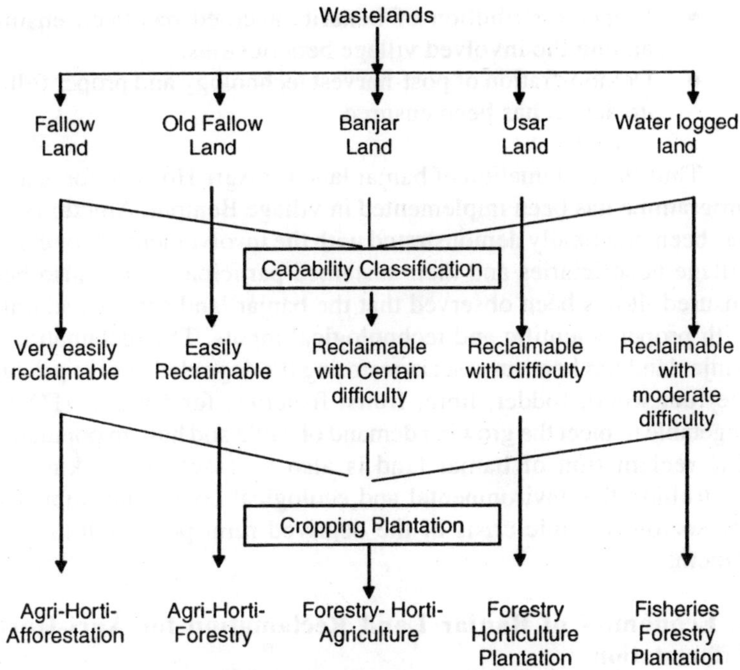

Capability Classification of Wastelands

management, mechanical processing, organic and inorganic amendments and micro-organism based bio-fertilizers and bio-pesticides.

- Nursery raising for various social forestry and horticultural plants has been planted.
- Green manuring has been ensured through Sanai/Dhaincha and agricultural cropping process has been practically demonstrated.
- Barley, Urad, Mung etc. agricultural crops have been cropped in the reclaimed Banjar land.
- The bio-scientific management of multi-intercropping on the reclaimed banjar land has been practically demonstrated.
- Proper scientific demonstration of harvesting techniques and recycling of crop waste for bio-fertilizers has been implemented with the involvement and sustained participation of the beneficiaries.

- Proper distribution of benefits accrued has been ensured among the involved village beneficiaries.
- Demonstration of post-harvest technology and proper follow up action has been ensured.

Thus the reclamation of banjar land for Agri-Horti-Afforestation programme has been implemented in village Benipur, Amethi Block has been practically demonstrated with the involvement of concerned village beneficiaries and their sustained participation has also been ensured. It has been observed that the banjar land can be reclaimed with proper scientific and technological inputs. The reclamation of banjar land has been an asset in restoring the degraded land to produce the fuel wood, fodder, fibre, fruits, fisheries, food grains (F^6) and vegetable to meet the growing demand of cattle and human population. The reclamation of banjar land is also an asset in checking and controlling the environmental and ecological crisis, and improving the socio-economic crisis of the deprived rural poor multitudes of Amethi.

6. Economics of Banjar Land Reclamation for Agri-Horti-Afforestation

The restoration of ecological and land degradation is the need of the hour for deprived rural poor because the explosive population growth tremendous deforestation, ecological and environmental degradation has brought us face to face for the scarcity of fuel wood, fodder, fibre, fruits, food grains, fisheries, vegetables and wood for the day-to-day requirements of the common people. The over-exploitation and excessive cultivation, high doses of fertilizer, canal irrigation are one of the basic seasons for the growth of wastelands has not been within the reach of the common farmers.

The adoption of various reclamation processes for wastelands, as accentuate soil erosion is not desirable for the long term point of view. In view of the above constraints, it is essential that pros and cons of wastelands reclamation are made amply clear to the farmers, so that they can adopt most appropriate remedial measures in the initial stages of reclamation tenure, subsequently the cultivators will be able to desire considerable benefit from the reclaimed land. The cultivators must be trained, motivated, to adopt various measures for the reclamation of wastelands only by disseminating to them the

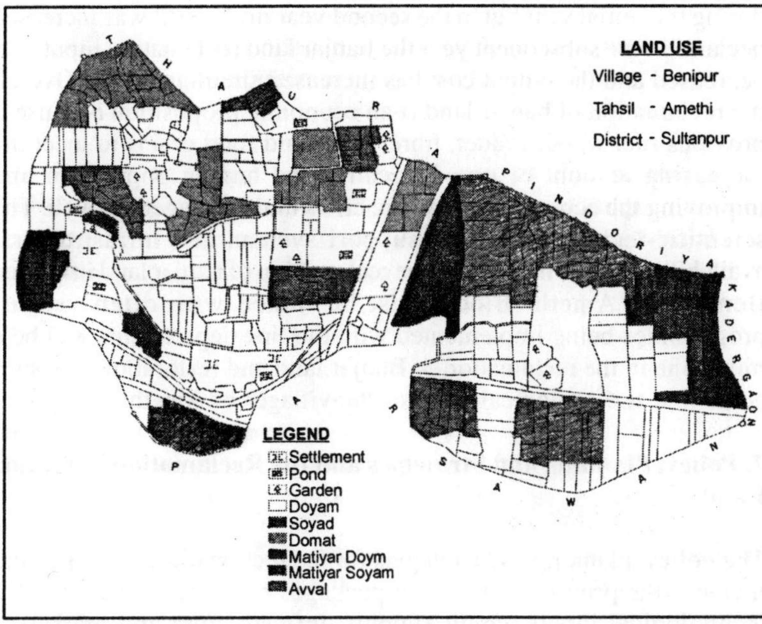

information regarding the economic benefits in the form of additional agricultural problems, which cultivated paves the way for socio-economic development and restoration of the ecological and environmental crisis of the villagers.

The banjar land reclaimable with certain difficulty has been implemented in the village Benipur of the Amethi Block. It has been recorded that a sum of Rs. 92,300 has been incurred for the banjar land development, nursery raising, Afforestation-Horticultural Plantation and agricultural cropping and implementation of post harvest techniques during the first year while a sum of Rs. 57,500 has been expended during the first year of the reclamation of banjar land, while Rs. 68,500 has been achieved as output cost from fuel wood, fodder, fibre and fruits during the first year and Rs. 82,000 has been achieved in the second year of the banjar land reclamation process. It has been noticed that in the first year the benefit was in minus but in the second year it increased and the farmer was benefited with Rs. 24500. Thus during the two subsequent years, very nominal benefit has been accrued from the banjar land reclamation.

It has been observed that the reclamation of banjar land for afforestation-horticulture-agricultural cropping was not beneficial during the initial year but in the second year the benefit was increased because in the subsequent year the banjar land reclamation input cost decreased and the output cost has increased simultaneously. Overall the reclamation of banjar land is an economic proposition because it provides fuel wood, fodder, fibre, fruits and food grains to meet the increasing amount of growing cattle and human population and improving the eco-socio-economic crisis of the villagers as well. The scientific-technical back up support system, and infrastructural availability to the villagers for the reclamation of banjar land in village Benipur of Amethi Block must be linked with often various programmes being implemented with the line departments will be a milestone in the reclamation of Banjar land and restoration of socio-economic and ecological crisis of the villages of Amethi.

7. Policy, Planning and Strategies and the Reclamation of Banjar Land

The policy, planning and strategies for the reclamation of banjar land includes the provision of several packages of suitable scientific and technological inputs, credit support, infrastructure and marketing support system.

Imaginative strategies for communication with the village beneficiaries should be systematized and extended to grassroots level beneficiaries. Appropriate use of mass media and audio-visual techniques should be made available to motivate the farmers for their enthusiastically involvement in the reclamation of wastelands. **"Greening of the wasteland"** with the active involvement of the deprived rural poor **people and people's participation** in the reclamation of banjar land for **Agri-Horti-Afforestation management** including incentives to the beneficiaries must be initiated and ensured.

Identification of wastelands, classification of wastelands and its distribution at village level should be analyzed, i.e. usar land, banjar land, old fallow land, fallow land, water logged land and other types of wastelands in the various villages of Amethi has been identified and its distribution has been also analysed.

Various natural and human factors responsible for the growth of wastelands has been explained at village level in Amethi Block of the District Sultanpur.

The genetics of the wastelands and the various determinant

responsible for the growth and formation of wastelands has been explained. The genetic of wasteland has been classified into three categories, viz. **marginally degarded land**, **partially degraded land** and **severely degraded lands**. The suitable strategies should be worked out for various types of wastelands based on genetic and formation of wastelands.

The capability classification of wasteland as per suitability of the soil to produce the crops has been analysed and strategies must be adopted to analyse the capability classification of wastelands based on the reclamation capability of the wastelands.

Testing of the soil, land preparation, levelling and bunding, leaching and flushing, green manuring, organic and appropriate irrigation management, bio-scientific management etc. should be adopted for the reclamation of wastelands.

Action oriented planning within a specific frame should be adopted for reclamation of wastelands, stepwise proper planning for the reclamation of wastelands starting from land preparation, soil testing, irrigation management, organic and inorganic amendments as per requirement of the soil for agricultural cropping and horticultural-afforestation programmes should be formulated and demonstrated to the needy poor beneficiaries of the villages.

Extension programmes for the reclamation of wastelands for agriculture horticulture-forestry-fisheries programmes should be practically demonstrated to the deprived resourceless poor beneficiaries.

Strategies for nursery raising including land preparation, polythene bag filling with mixed sand, soil, fertilizer, and seedlings and irrigation management and maintenance and management of nursery should be approximately demonstrated to the village beneficiaries.

Agricultural cropping for the Kharif, Rabi, Jaid cropping seasons especially paddy, wheat, barley, mung and urad and sanai/dhaincha etc. should be properly demonstrated on the reclaimed wastelands and agricultural cropping should be planned as capability classification of wastelands and fertility status of the soil.

Strategies for horticultural plantations including—Aonla, Papaya, Ber, Mango, Guvava, Karonda, Lemon, Banana etc. should be demonstrated on the reclaimed wastelands along with the leguminous trees besides being salt tolerant which helps in the fixation of nitrogen of the soil. The multi-inter-cropping along with horticultural plantations must be propagated at village level.

Afforestation strategies on the various types of reclaimed wastelands including plantation of Cineraria, Prosopis Juliflora, Azadi Rochta Indica, Tamfix Indica, Albizia Procera, Pongamia Pinnata, Eucalyptus Tereticorins, Shisham, Sagaun, Leucaena Lececacethala plants should be planted to produce fuel wood, fodder, fibre to meet the increasing need of the growing cattle and human population.

The waterlogged land should be developed for fisheries cultivation. Pond should be constructed in the waterlogged land in which fisheries culture may be developed and the bunds of the ponds should be utilized for horticultural and social forestry plantation with the view to restore the ecological imbalances.

Gram Banjar Bhoomi Vikas Pariyojana linking with NAREGA (National Rural Employment Guarantee Act) and allotment of the wastelands on patta/Tree Patta basis to the landless/agricultural labourers should be promoted and implemented with the involvement of village panchayat and representative of the local village panchayat for successful implementation of the reclamation of wastelands for the Agriculture-Horticulture-Forestry-Fisheries programme at village level. Greening of Gram Yojana should also be initiated at village level with the involvement of village Panchayat.

Compensatory fund for Afforestation (COMPA) for diversion of Forest land for non-forest use should be developed not only at the state level but also at District-Block-Village level to control the deforestation and use of forest land for non-forest purpose and compensate the deforestation through regeneration of the forest cover.

Panchayat Van Yojana should be launched at village level and intensification of agro-afforestation management on wasteland at village level should be propagated and implemented with the involvement of village beneficiaries for ecological restoration and eco-development and diversion of forest land for non-forest use and restriction on removal of fuel wood as head loads and availability of the alternative fuels.

Education, training for agro-afforestation management on wastelands for villagers and interested needy beneficiaries should be organised to meet the increasing demand, for fuel wood, fodder, fibre, fruits, fisheries and food grain (F^6) for growing cattle and human population and for restoration of ecological imbalance and improving the environmental and socio-economic crises of the villages for integrated development of the deprived rural multitudes.

8. Conclusion

The explosive population growth and over-exploitation socio-economic crisis, we have less than 11 per cent forest cover. The fuel wood requirement of our country is 130 million tonnes per year out of which only 39 million tonnes are fetched from the forest. There has been alarming challenge for loss of our agricultural land which is 175 million hectares and 40 million hectares of forest land. We are loosing 1.5 million hectares of forest land every year.

It has been observed that the wastelands are significantly distributed at village level in Amethi Block and Block level in Sultanpur District which has become a challenging problem to the planners and development agencies engaged at Sultanpur. The wastelands in Sultanpur, and Amethi Block has been left out of cultivation for very smaller reasons and ignorance of the farmers, it can be easily reclaimed adopting latest scientific and technical inputs.

The relationship between man and nature is symbiotic. The combination of natural and human factors plays an important role in the spatial distribution of wastelands. The *p*H value of the soil, land concentration, scheduled caste population and behaviour of wastelands are the dominant variables in explaining the variations in wastelands along with the other natural and human factors. The variables which features dominating in explaining the variations in wastelands are the natural factors, viz. *p*H value of the soil, drainage density, behaviour of water table and human factors which comprises mainly the land concentration and scheduled caste population. There are other natural and human factors responsible for the proportionate distribution of wasteland at village level in Amethi Block District Sultanpur.

The agro-afforestation in Amethi, Sultanpur has not been properly planned and appropriately implemented because it requires joint venture and efforts of the implementing agencies with the participation of local people. The agro-afforestation programme should be implemented for the people, by the people and demonstrated for the people and in the farmers field in an intergrated manners adopting latest developed and innovated scientific techniques.

Various policies and planning for agro-afforestation has been framed and implemented which could not provide significant impact at grassroots level. National Commission on Agriculture, National Wastelands Development Mission, National Forest Policy, compensatory fund for afforestation and Panchayat Van Yojana could not provide significant positive impact among the deprived rural poor who are the common sufferers of the deforestation and land degradation. The planning, the agro-afforesation management on wastelands at grassroots level should be planned and implemented with the involvement and participation of deprived rural poor beneficiaries.

The reclamation capability classification of wastelands has been analyzed as per the fertility status of the soil. The reclamation of Banjar land in Benipur of Amethi Block has been implemented and practically demonstrated. The agri-horti-forestry cultivation and plantation has been carried out through scientific and technological inputs. The reclamation of wastelands has been an asset in controlling the ecological imbalances and improving the socio-economic crisis of the deprived rural poor villagers of Amethi.

The economics of Banjar land reclamation for Agriculture-Horticulture-Forestry has been analysed. It has been observed that in the initial stages of banjar land has been comparatively less beneficial because the input cost was high while output cost was low but in the subsequent years of the reclamation process the input cost was lower and output cost was higher and the net profit also increased.

The strategies formulation and adoption of various measures for the reclamation of wastelands included the provision of the several packages of most suitable technological and scientific inputs, credit support, building up of a strong social infrastructure and arranging marketing system. The strategies for the reclamation of wastelands with the involvement of rural poor villagers for agro-afforestation management for appropriate exploitation of natural resources to meet the day-to-day requirements of the common people living in the remote rural villages.

Action oriented planning within a specified time-frame should be adopted for reclamation management. Strategies for land preparation, nursery raising, agricultural cropping, horticultural and forestry plantation should be properly planned and implemented. Gram Banjar Bhoomi Vikas Yojana Compensatory Fund for Afforestation for diversion of forest land and non-forest use Panchayat Van Yojana and intensification of agro-afforestation management of wastelands at village level should be implemented with the involvement of villagers.

It has been observed that the Amethi Block and District Sultanpur has geographic and historic importance. Various types of wastelands are significantly distributed in Amethi Block and District Sultanpur. Various natural and human factors a mainly responsible for the formation of wastelands. Agricultural cropping horticultural plantation and afforestation programmes are insignificantly implemented needs of the local people. The reclamation of wastelands can be easily implemented through land preparation, appropriate management of irrigation system, proper application of organic and in organic amendments, and other activities to restore the fertility status of soil for agricultural cropping, horticultural and forestry plantation and fisheries programmes as per suitability of the land. Several policies and planning have been formulated for the reclamation of wastelands including Wastelands Development Mission and National Forest Policy impact and message among the common people at village level and it was limited within the official files at National and State levels and at the maximum extent up to District level but it could not reach up to the field, i.e. grassroots level or at village level for whom the policy and planning for agro-afforestation management has been formulated.

The Hon'ble Supreme Court of India has also concern and instructed the Government to develop Compensatory Fund for Afforestation (COMPA) for the diversion of forest land to non-forest use but it is not being implemented at village level successfully.

The Panchayat Van Yojana, Gram Banjar Bhoomi Vikas Yojana, Greening of Wastelands Programmes must be launched at village level with the sustainable involvement of the Village Panchayat and interested needy villagers. Agri-Horti- Forestry-Fisheries programmes should be implemented as per the requirement of the people, by the people, with the people and for the people. The common land/waste lands must be allotted on Patta basis or Tree Patta basis to the landless/ agricultural labourers and the reclamation of such wastelands should be linked with the 100 days employment scheme of Mahatma Gandhi National Rural Employment Guarantee Act (MNREGA) and requisite scientific and technical backup support should be made available not only for the forestry-fisheries cropping and plantation programmes respectively.

It has been observed that the economics of wastelands reclamation is beneficial viable proposition to the villages beneficiaries. Initially the benefit was less because the input cost was very high and output cost was low but from the second and the subsequent years of the wastelands reclamation input cost has been reduced and the output cost was increased, hence the reclamation of wastelands for Agri-Horti-Forestry-Fisheries programmes has been beneficial to the villagers. No doubt it has been helpful in restoring the ecological, environmental crisis and restoration of the degraded lands and improving the ecological imbalances of the area and in improving the socio-economics crisis of the deprived poor villagers on the one hand and it has been a mile stone in fulfilling the increasing demand of growing cattle and human population for the fuel wood, fodder, fibre, fruits, food and grains, fisheries etc. on the other. Thus the reclamation of wastelands for Agri-Horti-Afforestation and Fisheries programme as per suitability and capability of wastelands is an asset for integrated rural development and eradicating poverty of the resourceless villagers, generating income among the rural poor masses and also increasing the income of the deprived rural multitudes.

6

Resource Utilization of Sundarban Mangrove Forest People: An Ecological Perspective

Mohammad Majumdar, S.M. Monirul Hassan and *Rashed-Un-Nabi*

Introduction

The exploitation of resources by a community of people is always sensitive to the existing environment. Individuals are rather consciously exploiting resources from their surrounding environment. People and environment do not merely exist side by side for their own respective survival. On the contrary, there is a continuous mutual interaction between people and its immediate environment, as if they exist for each other's sake. Rational individuals, with their rational mind exploit resources in the manner that the existence of surrounding nature is not jeopardized. Any harm done to the nature ultimately unsettles the mere existence of human society itself. So, whatever actions of a community of people taken in exploiting resources, they always take a good care of over-exploitation of the nature. Based on the research data from Sundarban mangrove community, the authors of this paper argue that despite the presence of maximizing resource utilization motivation, people are not indifferent to the destruction of their immediate environment. Knowing the critical balance between man and nature people of Sundarban area are involved in creating imbalance in it.

The authors will also argue that although seemingly, the people of Sundarban mangrove forest are exploiting resources extensively, the people themselves are aware of the action and consciously taking measures not to disturb the carrying capacity of the current ecological niches. Ecological niches consist of people and organism and its physical environment, upon which all the organism including human being is dependent. As Hardesty writes, "For human or any other consumer organism that must "feed" for a living, the available energy and nutrients can be viewed as a variety of food resources in the form of biomass or materials. The resources vary in size, colour, distribution in space and time, temperature, mobility, and so forth. Those variants upon which an organism feeds is its niche. As because of continuous interdependence between men, animal and physical environment for their respective survival, each is committed to the maintenance of critical balance of the ecological niche.

Keeping this in mind, the authors of this article finally put forth two basic arguments, which would be followed in the rest of the writings.

Argument 1: People in the ecological niche are rationally exploiting resources for their own survival.

Argument 2: In ensuring their survival, people are also committing to the destruction of critical balance existing within a specific ecological niche.

Theoretical Approaches

Although ecological approach in social science is a recent phenomenon, its origin could be traced back long in the history. Only the study of understanding relationships between organism and environment systematically started emerging during twentieth century. "Anthropogeography" or "Possibilism" emerged in the light of highlighting mutual dependence of organism and environment. Gradually, ecological approach led the possibility of more penetrating form of analysis of studying the interaction of man with the rest of nature then either of these provide.

The anthropogeographic approaches amongst which the climatologically theory of Ellswarth Huntington was most notable one. According to this theory, human culture was shaped by environmental condition (Hantington, 1995). This theory did not incorporate the total view of environmental determinism. The proponent of this theory also admitted that human culture sometimes exist independent of geographic forces. The variations, which they faced contrary to this position, were accepted as an exception or accident. It is in this way the anthropogeographers explained the human culture, which were opposite to their theoretical position.

In the possibilist approach, on the other hand, the environment was seen not as causative but as merely limiting or selective force. Geographical factors, according to them, did not shape human culture—but this set boundary to the forms it could take at any place and time. Culture rather has taken place due to specific historical process. A.L. Kroeber's famous study of maize growing society of the North American aboriginal resembles this kind of theoretical approach. What Kroeber pointed out is that the maize growing of the aboriginal is not affected by environment, but the non-growing of maize production is influenced by the environment.

Neither of these views is simply wrong. Yet both are inadequate for a correct analysis. The anthropogeographers do argue that geographic factors play a dynamic, not merely a passive role in the development of human culture. Possibilist, on the other hand, argue that, environment cannot work as a causal factor in shaping human culture. The indeterminacy on either side here actually seems from a serious conceptual defect the two approaches share. Both initially separate the works of man and the processes of nature into different spheres—"Culture" and "Environment"—and then attempt subsequently to see how these two spheres externally affect each other.

The ecological approach attempts to incorporate human activities, biological transactions, and physical processes in one unified system, i.e. ecosystem. What is this ecosystem? ".... an ecosystem consists of biotic community of interrelated organisms together with their common habitat and can range in size, scope, and durability from a drop of pond water together with the micro-organisms which live within it on the entire earth with all of its plants and animal inhabitants. The concept of an ecosystem thus stresses the material interdependence among the group of organisms which form a community and the relevant physical features of the setting in which they are found. The task of the researches is to reveal the internal dynamics of such systems and the ways in which they develop and change. In the language of Paul Sears, "When the ecologist enters a field or meadow, he sees not what is there but what is happening there."

What is happening there is a patterned interchange of energy among the various components of ecosystem. The living things take material as food from their surroundings and discharge material back into these surroundings as waste product. This is a process going on ceaselessly and thus the organism and non-organism keep their mutual function and continue their existences. This process is referred to "The balance of nature". An example could be drawn in this respect. "A flock of sheep in a pasture, the sheep are, with their sharp, close-cropping teeth, apparently destroying the grass by ingesting it. But the sheep are also fertilizing the pasture with their manure. Thus, if the sheep were removed the pasture would, at least in many cases, be removed too; for trees would begin to seed and grow, finally killing off the pasture grass and where once was a field would now be a wood. The sheep and the pasture form an integrated, equilibrated system, each of them dependent upon the other for its existence. Such equilibria are commonly, of course, quite complex consider the neat balance between water, oxygen, light, heat, green plants, microscopic animals, insects and fishery in a pond."

Julian Steward develops this type of theoretical perspective by emphasizing the nature-culture interrelationship on specific issues. Steward, although, assert that all aspects of culture are in some indeterminate way, functionally interrelated, argued that the degree and kind of interrelationship is not the same in all aspects of culture, but varies. According to him, some aspects of culture are directly interrelated with nature and determine the characteristic of human society. This is called cultural core, while some other aspects of culture are not directly interrelated to specific environment, which he called 'the rest of culture'. Julian Steward maintained that two societies living in the same environment varies in the rest of the culture dimension, but they converge on the cultural core issues.

The main characteristic features of these ecological theorists are that both organism (biotic community) and inorganic elements are constantly involving in energy transformation for each other's existence, thereby, maintaining order in nature. Human culture is created in order to adapt to that kind of situation. So, what is happening or occurring in human society is not happening without the interference of environment and individuals of a society. They are, therefore, consciously or unconsciously acting in order to confirm this situation.

Methods of the Study

In this study, a proper methodology is needed in order to reflect our theoretical orientations. With this end in view we have selected our site.

Selection of the Study Sites

The study was carried out in two different sites of two different districts belonging to Sundarban Mangrove Forest (SMF) to compare the social-ecological dynamics of fishermen community of Bangladesh. Bamia Nil Dumor and Joymunir Gul village would be called site 1 and site 2 respectively. The distance between the two sites is about 100 km. The following table would give us a general overview of the study sites:

Subject	*Site 1 (Bamia Nil Dumor)*	*Site 2 (Joymunir Gul)*
Location	Union-Burigoalini Upazila-Shyamnagar District-Satkhira	Union-Chila Upazila-Mongla District-Bagerhat
Main River	Chuna	Passur
Inhabitants	Five thousnad (approx)	Four thousand (approx)
Main occupation	Fishing	Fishing
Fishing type	Artisanal	Artisanal

Brief Description of Site 1

The Bamia Nil Dumor village is divided into four Para (area): Dali Para, Gazipara, Molla Para and Sarkar Para. They are situated at the last southern part of the SMF under Burigoalini Union of Satkhira district. Approximately 2650004 people live in this upazilla. Among those 5.5 per cent are engaged in fishing. About 5 thousand people live in this village and most of them are artisanal fishermen.

Brief Description of Site 2

The Joymunir Gul village of the Chila union was established about 120-150 years ago. It is situated at the southern part of the Mongla Upazilla under the district of Bagerhat. Mongla has a population of

137947. Pasur river is flowing at the western side of the village and finally flow down to the Bay of Bengal. More than four thousand people live in this village and fishing is the main occupation of the villagers.

Preliminary Survey

A preliminary survey of the study was done during the month of September 2007 to assess the prevailing condition at the sites and issues to include in the study. This preliminary survey was conducted in the villages to appraise the nature and categories of fishing community activities to outline the questionnaire. The proposed two sites were selected to observe the social ecological dynamics of the artisanal fisherman community. During the preliminary survey it was found that residents in the study areas mainly live on fishing in the nearby rivers and used mangrove resources of the Sundarban Mangrove Forest (SMF).

Data Collection, Design and Sampling Procedure

The study was based on primary as well as secondary data. Data were collected between October 2007 and February 2008. Before preparing final questionnaire a draft questionnaire was developed keeping in view the specific objectives of the study. The draft questionnaire was pre-tested in the study area interviewing a few fishermen by the researchers themselves. The final questionnaire was then developed in the logical sequence so that the fishermen could answer chronologically. Questions related to socio-ecological condition, occupation, fishing, and income of the fishermen in the past and present as well as of family members were included. The interview schedule contained both closed and open-ended questions. The questionnaire was constructed in English. However, the questionnaire was translated into Bengali by the researchers and assistants throughout the interview processes.

To ensure a comprehensive study of the stated village fishermen community and to consider the time as well as expenditure, the researcher and the assistants together interviewed 100 fishermen from each site. The researchers engaged three assistants during the whole data collection time and interview. The assistants are the graduates of University of Chittagong, Bangladesh. Necessary instructions were given to the assistant by the researchers before the data collection and interview. Altogether 15 days (15.12.2007-31.12.2007) were required to collect data, interview and organized Focus Group Discussion (FGD). In the first stage, SOSILON, a local NGO provided household information of site 1 and WORLD VISION, another local NGO of site 2 from their previous study. Based on information, the researchers applied random data collection techniques to gather information from the villages. During the data collection period, the researchers and the assistants visited the respondent's house. Men as well as women were interviewed in both selected sites. Maximum of one and half hour was spent for each interview. If respondents were not willing to be interviewed due to their involvement in various activities or for some other reason, then the researcher went on to the next house to save time. Prior consent was taken from the male family head to interview the women in both study sites.

Secondary data were collected from relevant studies conducted by various educational and research organizations. Reports and published articles by some local NGOs which were working with the small scale fishers of coastal area of Bangladesh, and some newspaper reports were also used as secondary data. All these relevant documents as reports might be of use for data enrichment and analysis.

Identification of Stakeholders and Nature of Interaction

Sundarban Mangrove Forest has immense importance to the local people for their livelihood activities. Different stakeholders have emerged in this regard. Stakeholders have been identified in the following way:

Local user community—Artisanal fishermen, shrimp fry collectors, *bwali* (wood) collectors, *mwali* (honey) collectors and *golpata* (leaf) collectors belong to this category of people. People of these communities live in and around mangrove habitats and they had common interests on the mangrove resources, directly using the resource like fishes and other animals as food, wood as fuel, leaves and young shoots as fodder.

Local community—People of this community have indirect interest on mangroves, as they are not dependant on any particular mangrove resources. But they live within mangroves and get protections against cyclone, seawater intrusion etc.

Remote user communities—These people live far away from the mangrove forest but use resources like swamps, shrimp fry collected by local people. They are involved with aquaculture activities.

Government agencies—The Forest Department and the Fisheries Department are the two main stakeholders in the study areas to manage and protect mangrove resources. They implement the rules and regulation through direct and indirect interacting with the local fishermen communities. During the FGD, the respondents stated that Forest Department officials took bribe from the fishermen and others for their illegal activities and allow them to catch fish by using illegal nets and gears.

Supporters of Mangrove user Communities—In general, non-governmental organisations (NGOs) and voluntary agencies are the supporters of mangrove user communities. However, moneylenders (*dadondars*) also fall in this category as they provide loan to the fishermen for buying fishing equipments. SOSILON, GONOMUKTI, UTTARON, ASA, BRAC and ACTION AID are the prominent NGOs in the study area Bamial Nil Dumor. In contrast, BRAC, BRDB, MATSHAJIVI SAMITY, SANDANI LIFE INSURANCE and POPULAR LIFE INSURANCE were the prominent NGOs in Joymunir Gul. Micro credit is the main activities of the NGOs in the study areas. However, respondents stated that they obtained financial and technical help from the NGOs during crucial time. Some NGOs are also reported to be using these communities for collecting donation from the foreign organization.

All the stakeholders of the mangrove forest are involved in mutual interaction. Mainly cooperation and conflict are visible in their pattern of interaction. Fishermen sometimes sharing nets, boats, fishing knowledge, news of each other's family, sharing food, protects collectively from dacoit etc. They also make quarrel with other fishermen in setting nets and in utilizing other resources from the mangrove. Moneylenders are cooperating by lending money and giving marketing facilities for the fishermen. They are also involved in conflict by constraining fishermen to sell fish on lower price, cheating and so on. The forest department cooperates with fishermen by providing security and cutting of wood and fishing illegally. The forest department officials take more fees from the fishermen and take bribe. Thereby create conflicting situations between each other. What is observed out of the interaction pattern is that, the fishermen are the most oppressed of the whole lot and possess very little command at their disposal. In other words, although cooperation is taking place, conflict is a must due to unequal distribution of wealth, status and power among the stakeholders.

Demographic Description of the Community

The demographic distributions of the both sites (study area) are identical. Most of the respondents are male and very few female household heads. 83 per cent and 93 per cent are male household heads in both sites. Only 17 per cent female household heads are from site 1 and 7 per cent from site 2. More than 90 per cent respondents from both site areas married. Only 3 per cent and 8 per cent respondents of site 1 and site 2 are unmarried. More than 70 per cent of the respondents from both sites are aged not more than 50. In other words, from the above description it is clear that most of the respondents of both sites are in their maximum productive years and are involved actively in their livelihood activities. The following table illustrates the personal profile of household heads in the study areas:

Variables	*% in site 1*	*% in site 2*
Male	83	93
Female	17	7
Married	94	90
Unmarried	3	8
Widow	3	2
Age class (20-30 years)	21	24
Age class (31-40 years)	32	37
Age class (41-50 years)	13	26
Age class (50+)	34	12

The understanding of educational level of a given community is important to get a clear picture of the people. Five categories are used to define the education levels of household heads. (1) Can sign, (2) Primary (from 1st to 5th class education), (3) High School (6th to 10th class educatory, (4) S.S.C. (Secondary School Certificate), (5) Illiterate. In Bangladesh if some one only knows how to write his name, he/she is treated a literate. From the results of all respondents in both studied area, 29 per cent and 25 per cent are illiterate respectively in site 1 and site 2. Among the remaining respondents

43 per cent (site 1) and 23 per cent (site 2) can sign. Only 19 per cent of the respondents in site 2 had high school education. In case of S.S.C. level education, it is 10 per cent in the study area 2 and 0 per cent in the study area 1.

It would be interesting to see the nature of family size of the two studied sites. Two types of family (nuclear and extended) are found in both study areas. Nuclear type of family is the family where sons and daughters are separated from the parents after marriage and start their own family. On the other hand, extender family constitutes parents, sons and daughters live in the same household even after the marriage of the children. This type of family pattern is more often seen when children are young.

It is interesting to note that nuclear type family is dominant in both study areas being 55 per cent in site 1 and 72 per cent in site 2 respectively. In the extended family, most of the members go for fishing and they can extract more resources from nearby rivers as well as from the mangrove forest. It might be the reason for higher presence of extended family in site 1. On the contrary, single type family fails to extract much resource from the forest owing to few members.

Nature of housing pattern is also important to grasp the livelihood activities of any community. The majority of the houses in both study areas are katcha/Golpata (Nypa fruticans leaf) produced. Its construction materials are found from mud and mangrove. The lower part of the house constructed with mud and the upper part with Golpata as roof thatches, usually the roof are covered by Golpata. It is highest 66 per cent in site 1 and 56 per cent in site 2. It showed that the respondents use a huge amount of mangrove materials for household construction. The following table would show this:

Variables (House Type)	*% in site 1*	*% in site 2*
Katcha/Golpata (Mangrove leaf)	66	56
Tin	12	8
Bamboo	22	36

Resource Utilization Strategy and Its Sources

As it is mentioned, fishing is the principal occupation of the research area. There are other secondary occupations utilized by the local community. In doing so, the livelihood activities of the local community take away its pressure on fishing and keeping the carrying capacity of the mangrove areas. Secondary occupation consists of agriculture, crab collection, fry collections, golpata cutting, fuel wood collection, boat making, wage labour, netting etc. The purpose of taking over secondary occupations is to supplement the shortage of income coming from primary occupational activities. The majority of the respondents in site 1 (40%) had an average monthly income ranging 3100 to 4000 taka from main occupation. In contrast, 25 per cent respondents in both study sites had an average monthly income of 4100-5000 taka (100 Taka = 1£).

The resource utilization strategy of the respondents could be grasped by seasonal calendar analysis. Seasonal calendars represent the activities of a community over a year. People undertake very different activities as the season change. The same is true for the artisinal fishermen communities of the study area. The fluctuation of key resources such as fishes as well as mangroves through the year compels the fishermen to take different activities in different times of the year. Four seasonal calendars, two for each study area are constructed. Calendars are organized according to Bengali calendar month to gather information on fishery resources and mangrove resources based on the economic importance as well as the availability of the resources throughout the year. Ten fishery resources are (species) Faisa, Datin, Kain, Illish, Tengra, Vetki, Shrimp (Golda and Bagda), Chitra, Shoal and Rui. On the other hand, ten mangrove resources are fish, golpata (mangrove leaf), honey, nolkata (fire wood), shrimp fry (Golda and Bagda), Singra kat (fire wood), Gorgan (mangrove tree), Sundori (mangrove tree) and crab. According to the respondent's statement, resource availability differs between study sites. For example, quantity and availability of the fish species is high in the Bengali month of Asar, Srabon and Bhadra (June-August). However, the availability of Faisa is high in the month of Baishak, Jaistho and Ashar (April-June) in the study area 2. Mangrove resource availability and extractions are similar in both study areas. When the availability of resource is high, the respondents of the study areas also increase the extraction of it. It should be mentioned that a government permission of resource collection is required prior to the move. However, by giving bribe to the Forest Department officials, illegal extraction of the mangrove resources is carried out by the people.

Pattern of Fishing Activities

As fishing is the principle source of resource to the respondents, it is necessary to know the pattern of fishing activities of these people. Fishing with partnership is best desirable practice of the fisherman. 79 per cent people in site 1 and 88 per cent of site 2 are involved in partnership fishing. Generally 3-4 people catch fish together in partnership mode. Neighbours, friends, relatives or family members of the same community group together for partnership fishing. The purpose of making partnership fishing is to earn more within a short time. Besides, protection from tiger attacks requires more than one person; usually three for fishing inside Sunderban. One maintains the net; one controls the rudder of the boat and the third one serve as a watchman against tiger movement. Every year, a large number of people become the prey of tiger attack. So partnership fishing is not only economically viable but also desirable in ensuring security for people themselves. Women and children who are engaged with fish fry collection do it individually. They fail to go for fishing in groups because of social and religious constraints.

Description of Fishing Areas and Fishing Boat

Artisanal fishing communities in both study areas are involved with inshore fishing particularly for shrimp fry collection by using set bag net and drag net. 'Set bag' net is usually used by male fishers inside the SMF and 'drag net' by women and children fishers. Every family has got one to three 'drag nets' and one 'set bag net'. Sometimes fishermen hire the net from the local dealers or traders. However, fishermen in sites 1 and 2 use the same type of fishing nets. Most commonly used nets are: *nylon jal, basan jal* (gill net), *jaki jal* (cast net), *panamara jal* (Push net with set bag net characters), *keoa jal, tana jal* (push net or drag net). The majority of the fishermen in both study sites have fishing nets. The total number of nets varies between the two sites. In site 2, 56 per cent of the respondents have more than one net. On the other hand, 48 per cent respondents in site 1, have 3-4 nets. More nets give the opportunity to catch more fishes in the nearby rivers and mangroves wet areas.

Artisanal fishermen in both studied sites use non-mechanized boats for fishing. They use good quality of mangrove wood for the construction of boats. Some fishermen are also engaged in boat making as a secondary occupation. This pattern has been revealed while talking to the respondents in group discussion. The majority of respondents have 1-2 boats for fishery and it is highest (64%) in site 1 and 61 per cent in site 2. Only 18 per cent and 11 per cent respondents have 3-4 boats in site 2 and 1 respectively.

Access to fishing ground in the coastal communities is mainly depended on fishing boats. This seems to indicate that many fishers have no access to these essential assets for their income. 25 per cent and 21 per cent respondents of site 1 and 2 respectively are deprived of fishing due to this cause.

Ecological Niche and the Fishermen

The ecological niche of the studied area has been so far discussed. The fishermen out there are living by exploiting the particular ecological niche, i.e. Sunderban Mangrove Forest (SMF). The fishermen are not only depending on SMF for primary livelihood activities, they are also using it as secondary livelihood activities. In this way, the fishermen are maintaining their existence generation after generation. Thus, the cycle of energy transformation is going on in that particular ecological niche (SMF). It is a fact that the equilibrium of a certain ecological niche could be maintained as long as, the carrying capacity of the niche is not exhausted. And the rational individuals, living in particular ecological niche always try to maintain niche balance. Otherwise they know this would jeopardize their own existence. The delicate balance between human being and nature is the precondition for the continuity of any social life. The community of the studied area has been so far successfully doing so.

The contribution of mangrove fisheries in SMF over last few decades has gone through two distinct phases; the first phase showed an increasing trend until 1988-89 and after that second decreasing phase started and still continues.

The present study shows that the equilibrium of the SMF has started dwindling. Various factors are responsible for this. Amongst them, population pressure is the most severe one. This population pressure is leading to over exploitation of mangrove, conversion of mangroves to settlement, rice fields, salt beds, and tourist resorts building etc. Besides, cleaning for timber and wood-chips industry and excessive extraction of timber for firewood are also the result of population pressure. This has been disrupting the carrying capacity of the SMF. The inhabitants of the community who are using mangrove forest are not ignorant of this situation. They are coping themselves with the present crises along with various measures.

First of all, the informants of the community prefer to go for fishing with collective effort. This study shows that people of both sites prefer collective effort during fishing. 79 per cent informants of site 1 and 88 per cent of site 2 prefer partnership fishing. Partnership fishing, i.e. collective fishing consist of 3 to 4 people for an efficient fishing.

Occupational diversity is another kind of strategy in coping with the odds of the niche. Fishermen are not only involved in fishing, they are also engaged in various activities such as agriculture, crab collection, fry collection, *golpata* cutting, fuel wood collection, wage labour, boat making, netting and so on. The following is the statistics of the people of both sites involved in secondary occupation.

Variables	*% in site 1*	*% in site 2*
Agriculture	10	16
Crab Collection	16	13
Fry Collection	15	18
Golpata Cutting	13	20
Fuel Wood Collection	17	15
Labour	13	10
Boat Making	8	2
Netting	3	2

Over-exploitation is the second strategy by means of which the fishermen are coping with the difficult situation. Over exploitation is unavoidable, even though the respondents are in principle, against it. What they are trying to reveal is that, the family size has become large; the resource is not enough to meet the demand. Therefore, despite the awareness of the negative impact of over-exploitation, the fishermen are continuously guilty of doing this action. This could be revealed in a statement given by a crab collector of site 1.

> "We are 7 family members. Earlier, when I caught crabs I always let juvenile crabs free. But now sizable crabs are dwindling. My average income is 195 Tk/day (2 £). Now I don't allow young crabs back to water. I retain them and thus supplement our daily food. We know we are destroying our future livelihoods. But we are helpless."

Excessive resource exploitation is not only happening due to population pressure, it has some other causes too. Increasing overhead cost of fishing, the informants think, has constraint people to think of over production. The reasons of increase overhead cost of fishing are extortion, bribe demanded by government officials who are entrusted to preserve ecological balance of SEM areas. During the group discussion in site 2 one of the informant's statements reveals this:

> "We are fishermen in the Chadpai range of Sundarban. We have to collect a permit for fishing by paying a certain amount. But permit issuing fisheries official always charge extra money. Our income for fishing per week is about 1000 Taka (£10). So we have some saving. But in Sundarban area there are a number of active illegal elements. They divide the total area into different zones for collecting extortion. We have to pay 500 Taka (£2) per week as extortion. Moreover some unscrupulous forest officials also charge extra money during fishing. In this way all of our income goes to the pockets of others. But we have to survive. So we compensate our loss by over production. We put more fish than the permit allows us under the duck of our boat."

This trend could also be seen in other resource exploitation areas of Sundarban Mangrove Forest. Rational individuals are (consciously) involved in over-exploitation of other resources there. The respondents of the two sites have clearly mentioned their conscious way of over exploiting resources. The following table would give us detail of the matter. The respondents were asked to identify the causes of deforestation. They answered in the following way:

Variables	*% in site 1*	*% in site 2*
Firewood	25	22
Illegal Cutting	29	30
Household consumption	35	33
No idea	11	15

Very few respondents, i.e. 11 per cent and 15 per cent of site 1 and 2 respectively told that they have no idea of deforestation process in Sundarban. In other words, the over-exploitation of resources is occurring not without the awareness of concerned people. Rather, they are well aware of it. They are also aware of current dwindling process of resources compare to the previous period. One respondent in site 2 expressed his idea as:

> "Twenty years before if I went for fishing in the Chila river, I could catch more than 5 kg different types of fish within 6 hours. Besides fishing, sometimes I went to catch crabs also in nearby mangroves. I was happy with the amount of catch but now I am not getting fish that is sufficient for my livelihood. I don't know why it is happening and how I will survive in near future. I might change my occupation rather than continue fishing."

Coping Strategy

The statement clearly shows that informants are well aware of the dwindling process of resources and the strategy to overcome this problem is to quit the present occupation. There are other means at their disposal in coping with the situation.

One of the most important means is working collectively. Family members are working together in order to withstand hardship they are facing because of the absence of adequate resource supply. Recruitment of people from outside the family needs extra monetary ability. This reduces the capacity of people for subsistence. In most cases, fishermen and crews are thereby recruited from the same family. When the father grows old, usually a son occupies the position of the skipper giving the father some easy job on land like making nets or mending it. Cooperation within the family is one of the most important strategies of the mangrove forest area. This puts constraint on the people to maintain family bond. In traditional Bangladesh, people create family through marriage. This is also there in our studied area. The frequency of remained married is very high in both studied areas. 94 per cent and 90 per cent household heads of sites 1 and 2 respectively are married. Being married, gives opportunity to the people to remain together and act as a source of security against vulnerability.

Occupational diversification and geographical migration is another effective means in coping with the odds of the society. Although fishing is the main occupation in both study areas, fishermen earn extra money by collecting honey, wood collection, and *golpata* (mangrove leaf) cutting from SMF. However, when fish catch is not sufficient to survive, the fishermen change their main occupation for a short period of time. The alternative livelihoods are boat-making, netting, agriculture, wage labours etc. During the FGD, fishermen talked about migration. In the Sundarban area during the winter season many fishermen migrate to some nearby islands for fish drying activities. Only migrant fishermen inhabit in these islands. But migration does not always ensure better livelihoods. Multi-Species fisheries of Sundarban provide unique opportunities for fishermen to switch target species in different periods of the season.

Role of Women and Children

Despite the traditional concept that women are destined to live in private sphere the informants of the studied area have shown significant deviation from the traditional women's role. People are responding to the odd situation by allowing women to take part similar kinds of activities with the male counterpart for supplementing shortage of income of the family. Females are now involved in fish collection, marketing fish, catching seed, making and mending net. Access to micro credit has also allowed women to take part in various homestead activities. Some women in Sunderban area also earn money by raising poultry and other livestock. The fishermen family for maximum exploitation of resources also uses children now. All these are being done with the conscious bent of mind. The families are not sending their children for education rather they take the children with them during fishing expedition. This is done by the fishermen in keeping their existence with utmost calculative mind. One of the informants of site 2 confirms to this kind of attitude. The informant says:

> "I know to educate children is a good thing. Due to poverty I have to take my child for fishing because I need a helping hand for fishing. In that case, I have to pay wage even if I do not get any catch (crab). For example, today we (use my child and wife) 3 persons earned 200 Tk (about £2) for the whole day fishing. If I took another person (by sending my child to school) I had to pay at least 100 Tk (£1) for 1 day. Do you think it is possible to maintain my family of 6 members with 100 Tk (£1) for 1 day? Firstly, we have to survive then this question of education."

Conclusion

The biodiversity of Sundarban Mangrove Forest (SMF) has been threatened by human intervention. The delicates balance of the ecosystem is now endangered. Over exploitation of resources is the key to understanding this pattern. Population pressure is causing the local people to extract resources as much as possible. How do we explain this situation? Is it the people responsible for all these unfortunate situations? Could we blame the rational individuals in this matter? Our analysis shows that rational individuals are well aware of the consequences of over-exploitation of resources. They have adopted, in case of Sundarban, various strategies to off set the odd situation they are in. The answer probably lies in the realm of structure. The structural elements, for example, existing state structure, lack of occupational diversification prospect in the country, low organizational ability have been shown more responsible then the people of the community. Last but not the least, people of the studied community is also responsible in causing the present ecological crisis in Sundarban Mangrove Forest (SMF). Further studies are invited in illuminating more insights in this regard.

Section II
Innovation of Socio-Economic Institutional Infrastructure

Introduction

Socio-economic and Institutional Infrastructure plays an important role in the formulation of need based village development planning. H.R. Yadav has attempted to explain the innovation, information system for Social Infrastructure for village planning in Amethi. Attempt has been made in this study to explain the household, population, workers and various social/human infrastructure information which plays an important role in the policy planning for integrated village development planning of village panchayat Sarai Khema of Amethi Block.

Henrique de Barros and Shashi Kant Srivastava has studied impacts of petroleum Industry on fishing communities, having a case study of the Todos os Santos Bay, Bahia, Brazil.

H.R. Yadav has tried to explain the innovation, information system for economic infrastructure in village planning. Attempts has been made in this study to analyse the economic infrastructure including, land cultivation, land ownership, irrigation system and urea under agriculture production facilities which are the backbone for the initiation of policy, planning for the integrated village development.

Uttara Yadav and Shailendra Singh has analyzed the sustainable yogic farming having a social view because everything can wait but not the agriculture because India is facing a great challenge of degraded natural resources like soil, water plants, air and above all human beings.

H.R. Yadav has made an attempt to analyze the innovation, information system for institutional infrastructure for village development planning in Amethi. Attempt has been made in this study to explain the available historical, religious, administrative communication, drinking water, energy, medical health, agro-industries, health services, educational infrastructure etc. to ensure the need based village development planning process with involvement of local people in Amethi.

A.M. Shekh and Arun Patel has explained 'e-krishi kiran' Village Information System for Scientific farming with the objectives of analysis of soil, fertilizer usages, to suggest alternative cropping pattern based on moisture availability index and to provide soil health card to each farmer to guide them about soil conditions and cropping pattern/practices to be followed.

7

Innovation, Information System for Social Infrastructure in Village Planning

H.R. Yadav

1. Introduction

Plan formulation and implementation of development programmes are an intensive task. The management of data as available in the village-block and district level sources are not properly managed and it is not sufficient to meet the need base integrated village development planning. The traditional and conventional system of the collection and collation of data, storage are not amenable for quick updation, retrieval and holistic analysis approach system due to sectoral data-sets flow and lack of information on the quality of the required data.

Village level integrated development planning process requires a detailed knowledge of social/human infrastructure data and its interdependencies between various humans, i.e. social sectoral indicators to resolve often conflicting demand. It requires appropriate data management and analysis sectoral data on social infrastructure data base.

In view of the above, an attempt has been made in this study to analyze the households, population (male-female, caste composition, workers and of the social/human infrastructure data which play an important role in the policy planning for integrated village development of village Sarai Khema of Amethi Block.

2. Objectives

The main objectives of the study is to analyze the soil infrastructure especially population caste composition, work force and educational components of the Village Sarai Khema of the Amethi Block.

3. Social Infrastructure

Social aspects play an important role in the overall development of any area, because social indicators are one of the major determining factors in the policy formulation of need based planning of the village. The caste composition in the rural India is based on the occupational structure of the country. Caste composition also plays an important role not only in the planning process but also in the implementation of the planned development programmes.

Peoples participation in the formulation of planning process and their participation in the implementation of rural development programmes but it will be an asset for maintenance and sustainability of the complete rural development programmes. Therefore, an attempt has been made to analyze the suitable indicators related to the social sector mainly responsible for the integrated economic development. In view of the above household, population, workforce, educational profile and other social indicators have been analyzed through available collected data in the village Sarai Khema of Amethi Block.

4. Geographical Analysis

The total geographical area of the village Sarai Khema is 557 hectares. Village Panchayat Sarai Khema comprises three revenue villages, viz. Sarai Khema, Dadupur and Ram Nagar Mafi.

Thus it is evident that main Sarai Khema village comprise of Sarai Khema Khas, Munshiganj, Basuapur, Halapur, Benimadhav, Kutia and Ahmadnagar villages and Dadupur village is comprised Dadupur Khas, Pure Kutur and Jagapur revenue villages while Ram Nagar Mafi village is comprised Pure Pandit and Mainapur revenue villages. It is known fact that Sarai Khema is the main village panchayat while Sarai Khema, Dadpur and Ram Nagar Mafi are the revenue villages of the main village but these revenue villages are also comprising sub-revenue villages of the Amethi Block.

Village Panchayat
↓
Sarai Khema
↓
Revenue Villages

↓	↓	↓
Sarai Khema	Dadupur	Ram Nagar Mafi
↓	↓	↓
Sarai Khema Khas	Dadupur Khas	Pure Pandit
Munshiganj	Pure Kutur	Mainapur
Basuapur	Jafapur	
Halapur		
Benimadhav		
Kutia		
Ahmadnagar		

5. Population

The total population of the village Sarai Khema is 3814 out of which revenue village Sarai Khema constitutes 2238 population followed by 927 population in village Dadupur and 649 population in village Ram Nagar Mafi.

Population (Village Sarai Khema, Amethi)

Sl. No.	*Name of the Villages*	*Total Population*	*Male*	*Female*	*Children*
1.	Village Panchayat Sarai Khema	3814	1482	1378	954
1.1	Sarai Khema	2238	946	883	409
1.2	Dadupur	927	276	258	279
1.3	Ram Nagar Mafi	649	260	237	266

The main village Sarai Khema is having 2238 population out of which 946 male, 883 female and 409 children population. Revenue village Dadupur has total 927 population out of which 276 male population, 258 female population and 279 children are available in the village. There are total 649 population which contributes 260 male and 237 female population and 266 child population in Ram Nagar Mafi revenue sub-village of village Panchayat Sarai Khema of Amethi having nominal household and noted 70 Muslim household in the village Sarai Mafi of Amethi Block.

6. Households

There are 567 households in village Sarai Khema including its revenue villages, viz. Sarai Khema, Dadupur and Ram Nagar Mafi and other sub-villages.

Castewise No. of Households (Village Sarai Khema, Amethi)

S.No.	*Caste*	*No. of Households*
1.	Chamar	80
2.	Kori	80
3.	Barbar (Nai)	03
4.	Carpenter	05
5.	Blacksmith	01
6.	Washerman	07
7.	Yadav	25
8.	Brahmin	25
9.	Kshatriya/Rajput	35
10.	Kalwar	14
11.	Teli	01
12.	Maurya	25
13.	Bhuj	04
14.	Chauhan	50
15.	Kahar/Kashyap	60
16.	Muslim	70
17.	Dharikar	01
18.	Pal/Gadaria	12
19.	Hela/Sweeper	10
20.	Halwai	05
21.	Barai	01
22.	Verma/Kurmi	04
23.	Chamar Mangata	04
24.	Bahelia	45
25.	Kayastha/Srivastava	06
	Total	567

Thus there are 25 castes living in the village Sarai Khema and various revenue villages. Highest number of 80 households belong to Chamar caste (Scheduled Caste) and Kori (Scheduled Caste) 80 numbers of households are found in village Panchayat Sarai Khema followed by 70 Muslim households and 60 household belong to Kahar/Kashyap caste family. There are 50 Chauhan caste households. Thus there are 182 scheduled caste households in village Sarai Khema and 255 numbers of households belong to other backward class families in the village Sarai Khema. There are 25 Brahmin households followed by 35 Kshatriya households and 6 numbers of households belong to Kayastha/Srivastava families in village Sarai Khema of Amethi Block and there are 70 Muslim households in the village.

In view of the above, it is evident that proportionately 255 backward caste households are dominantly distributed in village Sarai Khema followed by Scheduled Caste households while very nominal 25 Brahmin, 35 Kshatriya and 6 Numbers of Srivastava/Kayastha household families are distributed in the village Sarai Khema.

7. Caste Composition

The caste composition of the village Sarai Khema is very diversified. There are 14 per cent Chamar caste and 14.1 per cent Kori caste which totally contributes 28.2 per cent of the population. The proportionate distribution of the Scheduled Caste population is 32.1 per cent to the total population of the village Sarai Khema in Amethi block.

Caste Composition: Village Sarai Khema, Amethi

S. No.	*Caste*	*Population in %*
1.	Brahmin	4.41%
2.	Thakur/Rajput	6.17%
3.	Scheduled Caste	32.1%
4.	Minorities	12.35 %
5.	Other Backward Castes	45.1%
6.	Kayastha/Srivastava	1.01%

The distribution of backward caste is very significantly distributed in village Sarai Khema, Dadupur and Ram Nagar Mafi villages. There backward caste population contributes 45.1 per cent of the total population while Kayastha/Srivastava (Upper Caste) population of the village is 1.01 per cent. The minorities population contributes 12.35 per cent of the total population, while upper caste Brahmin contributes 4.41 per cent and Kshatriya/Rajput contributes 6.17 per cent of the total population of the village. It is to be noted that Yadav backward caste constitutes 4.4 per cent, Chauhan caste 8.82 per cent, Kahar/Kashyap caste 10.6 per cent and Bahelia a backward caste constitutes 7.94 per cent of the total population of the village.

8. Workers

It is known fact that village Sarai Khema is agriculturally dominant village. More than 79 per cent population of the village Sarai Khema is engaged in the agricultural activities. There are 48.9 per cent cultivators in the village Sarai Khema of Amethi Block. There are 210 animal husbandry workers in village Sarai Khema out of which 81 animal husbandry workers are found in main revenue village Sarai Khema, 69 animal husbandry workers in the village Dadupur and 60 animal husbandry workers in village Ram Nagar Mafi. Most of the backward caste population having animal besides this they are also engaged in agricultural activities and other occupations for their livelihood. There are more than 79 per cent primary sector workers followed by secondary and tertiary sector workers.

There are 35 Mason workers in the village Sarai Khema and 7 workers are engaged in the transportation/communication activities. Approximately 15 persons of the village Sarai Khema are engaged in the educational institutions. Agricultural labourers are also significantly distributed in the village. It is estimated that more than 38 per cent population of the village are engaged as agricultural labourers in the village Sarai Khema of the Amethi Block. There are 7 washermen families in the villages and 4 Tailor Masters in the village Sarai Khema. It is known fact that most of the people of the village are engaged in their caste oriented activities.

9. Education

Literacy ratio of the village Sarai Khema is quite significant. There are 600 primary educated persons out of which 286 people are found in the revenue village Sarai Khema, 170 persons in Dadupur and 144 persons are primary educated in Ram Nagar Mafi.

There are 356 persons are having junior education out of which 155 junior educated persons are found in revenue sub-village Sarai Khema, 106 in Dadupur and 95 junior educated persons are found in Ram Nagar Mafi revenue sub-village. There are 265 high school educated persons and 145 intermediate educated persons in the village Sarai Khema of Amethi Block. There are 89 graduates and 40 Post graduate persons in the village Sarai Khema. There are 25 technically educated persons in the village Sarai Khema while only one graduate technical educated person is found in the village Sarai Khema of the

Education: Village Sarai Khema—Amethi

Sl. No.	*Education*	*Sarai Khema Total*	*Sarai Khema*	*Dadupur*	*Ram Nagar Mafi*
1.	Primary Education	600	286	170	144
2.	Junior Education	356	155	106	95
3.	High School	265	114	93	58
4.	Intermediate	145	78	42	25
5.	Graduate	89	51	22	16
6.	Post Gradaute	40	26	11	3
7.	Technical Education	25	15	6	4
8.	Graduate Technical Education	01	01	–	–

Amethi Block. The illiteracy ratio of the village Sarai Khema is negligible, while very old persons are not educated but the young generation is educated. Most of the Scheduled Caste and backward caste population is less educated who are having mainly primary, junior and high school educated in the village Sarai Khema. There is dropout for the higher education due to the socio-economic backwardness. There are noteworthy, i.e. significant ratio of the educated unemployed rural youths which has led towards economic backwardness of the village Sarai Khema of the Amethi Block.

10. Conclusion

Social structure and infrastructure of the village Sarai Khema and its revenue villages are one of the dominant factors in the integrated socio-economic development of the village Sarai Khema of the Amethi Block. It has been observed that cast composition which is based on the traditional occupation of the villagers plays an important role in the integrated development of the village Sarai Khema.

The total population of the village Sarai Khema is 3814 out of which 1482 male and 1378 female and 954 child population. Main Sarai Khema villages comprises 2238 population followed by 927 population in Dadupur and 649 population in village Ram Nagar Mafi of the Amethi Block.

There are 567 households in village Sarai Khema out of which 182 households belong to Scheduled Caste and 255 households belong to backward caste and 70 minorities household are found in the village Sarai Khema. There are 25 Brahmin households and 35 Kshatriya households while only 6 Kayastha households are found in village Sarai Khema of the Amethi Block.

There are 32.1 per cent Scheduled Caste population in village Sarai Khema while backward class population comprises 45.1 per cent the total population of the village. There are 4.41 per cent Brahmin and 6.17 per cent Kshatriya and 12.35 per cent minorities population in village Sarai Khema of Amethi Block.

There are 79 per cent population which is dependent on agricultural activities. The animal husbandry workers, Mason workers and other workers are nominally distributed in the village Sarai Khema. It has been observed that occupational structure of the villagers are dependent on the caste composition of the village Sarai Khema of the Amethi Block.

No doubt the village Sarai Khema is adjacent to the Hindustan Aeronautics Ltd., Munshiganj and Sanjay Gandhi Hospital, Munshiganj, while Munshiganj is also one of the revenue village of the village Panchayat Sarai Khema of the Amethi Block. There are 79 per cent primary sector workers followed by the secondary and tertiary workers in the village Sarai Khema of Amethi Block.

There are significant literarcy ratio in the village Sarai Khema. Most of the population of the villages are having primary, junior and high school education followed by Intermediate education. There are significant dropouts in the higher education due to socio-economic backwardness of the village.

It has been observed that most of the backward caste and Scheduled Caste population are less educated having only primary education and they are also having low size of land holding of which most of the land concentrated among 10.38 per cent population of the village while 89.42 population of the village are owners of the low size land holdings which is less than one hectare.

Thus the social composition of the village Sarai Khema plays an important role in the integrated development of the village. The formulation of the planning process must consider to involve the caste composition and occupational structure of the villages to ensure need based planning for the implementation of integrated rural development of the village Sarai Khema of the Amethi Block.

8
Impacts of Petroleum Industry on Fishing Communities: A Case Study of the Todos os Santos Bay, Bahia, Brazil

de Barros and *Shashi Kant Srivastava*

Introduction

Minimizing negative social and environmental impacts of oil production has been of growing concern for the industry and the general society over the last years. Taking as an example the region of the Todos os Santos Bay, close to Salvador, in the state of Bahia, Northeast Brazil, the pioneering character of the exploration techniques carried out for five decades are now deeply imprinted in both physical and social environments. In particular, local fishing communities have been reduced to an almost indigence, in sharp contrast with the power and resources displayed by the petroleum activities in the area.

From the late 1990s, the decommissioning of the Dom João Mar oil field has aggravated those conditions, as severe restrictions have been imposed to the artisanal fishing activities and the catching of mollusks and crustacean on the northeast of the Todos os Santos Bay. Added to other events provoked by the oil industry—in particular the launching of pipelines over fishing grounds—we perceive a perverse trend connecting the expansion of oil activities in the bay to the declining living conditions of professional fishers, men and women, in the region.

Fisheries have been generally acknowledged in recent literature sponsored by the oil industry as factors necessarily affected by activities of surveillance, perforation, and production of oil onshore and offshore. These impacts may vary from damages to natural habitats and the fleeing of species to the killing of critical rings of the aquatic food chain.[1] In spite of this recognition, recent proposals by the Brazilian industry have limited mitigation measures to the information to fishermen about the nature of the oil projects and the monitoring of impacts on affected fishing stocks (MME, 2007: 558-559).

Environmental Impact studies released during the last 10 years for the Todos os Santos Bay as well as the recent blueprint paper for the Brazilian petroleum industry (2007) do not propose adequate[2] compensations for fishermen communities affected by restrictions imposed to their professional activities. Also, contingency plans for the decommissioning of mature oil fields, such as the Dom Mar, entrenched well in the middle of the artisanal fishing grounds, have not come through.

In this paper, we discuss the most notable events produced by the oil industry since the starting of Brazilian oil exploration, early in the 1950s, and their impacts on working conditions of estuarine communities of the Todos os Santos Bay. As relations between industry and local fishermen have deteriorated over recent years, we propose an urgent revision of attitudes by industry to regain prestige and to promote a positive impact of oil exploration into the local depressed economies.

Fisheries in the Todos os Santos Bay

A field survey conducted from 2002-2004 has revealed the bleak conditions of the fishing activities in the oil production area. According to most of the fishermen interviewed, fishing conditions have deteriorated since the abandon of the oil field exploration, late in the 1990s. The apparent paradox is explained by the maintenance granted to the oil installations during normal operations and the fact that the oil rigs always works as artificial reefs to attract marine life. According to one of the local interviewees:

> "In my opinion, the Petrobras works has disgraced the canal; on the first years of fisheries there was no pollution; there was plenty

of fish and crustaceous. Today there are no more fish and crustaceous as used to before. Firstly, Petrobras caused the first damages. Drilling was performed with GM motors, too noisy. After drilling the hole, the 'base' got oil. The pipelines shaken during production; which fish wants to stay around? After abandon the oil field, Petrobras finished with us: fishing nets started to stuck in the iron wreckage."

The last part of the interview mentioned the impact of decommissioning of the oil field after it became unprofitable as overvalued Brazilian currency reduced imported oil prices late in the 1990s. By this time, Brazilian government was just starting to impose environmental licensing for oil activities. However, the decommissioning of an oil field such as the Dom Joao Mar was a new situation still to be included in the new regulations. More than 10 years later, licensing of decommissioning activities are still to be officialized.

As a result, the Dom Joao Mar decommissioning was carried on cutting the poles which supported the rigs just below sea level and sinking the aerial structures of seven concrete platforms and about 600 oil and gas wells. The remaining structures, consisting of iron pipes, together as well as the oil pipelines remain sunk or dangerously standing, hidden below sea surface. These old structures, although maintaining the function of attractors for marine life, have systematically damaged fishing nets as they stay hidden on murky waters with no signs to indicate their presence under sea level. This situation has remained for already 10 years (1997 to 2007) as Petrobras has offered no final solution for the clearance of the oil field debris. The remaining of the abandon oil structures on the sea bed contravene at least one IMO—International Marine Organization international convention signed by the Brazilian government in 1989, agreeing on the clearance of deccmmissioning of oil fields in waters less than 75 metres deep. According to the A-672 Resolution, in both offshore and continental waters,

"3.1 All abandoned or disused installations or structures standing in less than 75 m of water and weighing less than 4,000 tonnes in air, excluding the deck and superstructure, should be entirely removed."[3]

Claims that marine life is disappearing from the bay waters come from most communities of the region. Oil exploration, construction of two pipelines crossing the bay, a large oil terminal operating right on the middle of the bay, as well as a large oil refinery complex built in the early 1950s, are some of the events which, combined, are leading to the decay of fisheries and aggravating living conditions among local fisher, men and women.

A consequence of the demising of fisher profession has been the self-commiseration of the fishermen, regretting their sufferings and lamenting that urban people don't see them as the ones who work hard to feed the cities from the seas.

A condition which reinforces the hardships of fishermen from Sao Francisco do Conde and reduces their chances to overcome the cumulative impacts of oil activities is the technological stage of artisanal fishing in the region. Their canoes are primitively wood carved from a single tree log, which make them too unstable to reach areas beyond the limits of the oil field. This confines them to fishing very close to their communities, on the bay shores, while cleaner areas are available for those capable to meet the risks of crossing the bay far away from the production area. To date, since the starting of oil exploration, 60 years ago, no initiatives to improve technology among fishermen have surfaced.

The declining of sea and mangrove catches leaves few employment alternatives for families. A number of men and women have become part-time fishers, most of the time working on small professional jobs, such as plumbers, woodcrafters, housemaids etc. A political employment in the local municipality has also become an alternative for many people, the payroll funded by the oil royalties which turn Sao Francisco into one of largest budgets in the whole country. The municipal budget is in sad contrast with the destitutes of most of its population, especially those who depend on the sea. For most families there is no alternative then rely on retirement pension from the elders and small public subventions to families.

Housing among fishermen are precarious, 30 per cent of the interviewees residing in clay made ones, thatched with coconut leaves. For 40 per cent of them, there are no bathrooms, the necessities done in the woods around. Gas stoves are owned by 83 per cent, although daily cooking is done always in a charcoal/wood burner as gas has become increasingly expensive for them. Wood gathering has been mentioned by fishermen as an unavoidable threat to mangrove and the remaining native vegetation around communities.

The tragedy of commons' approach suggests that fishermen will not stop fishing while their comrades are still coming to the sea. In Hardin's perspective,

> (...) the only sensible course (...) to pursue is to add another animal to his herd. And another.... But this is the conclusion reached by each and every rational herdsman sharing a commons. Therein is the tragedy. Each man is locked into a system that compels him to increase his herd without limit — in a world that is limited. Ruin is the destination towards which all men rush, each pursuing his own best interest in a society that believes in the freedom of the commons. Freedom in a commons brings ruin to all (Hardin, 1968).

As much as productivity declines, more effort is applied by fishermen into the sea. Bombs are becoming more and more common despite widespread condemnation and the risk of imprisonment by authorities.

Fig. 8.1: Fishing Nets Used by Artisanal Fishers of the Todos Os Santos Bay (October, 2007)

Traditional artisanal fishing grounds affected by oil extraction or transportation include: (i) the bay waters, in special the areas closer to the old oil structures, turned into artificial reefs which attracted marine life over the last 50 years; (ii) sand banks, submersed in high tides, where mollusks are collected by women, and (iii) mangroves, traditional habitats of crustaceous and mollusks. Altogether, these places have been affected by oil spills, disturbances of the sea bed, drilling and their lubricating muds[4], and digging out sand banks to install pipelines.

All these events combined were likely responsible for fisheries decreasing productivity over the last 60 years. The following sections detail the activities of the oil industry in the region and their impacts on fishing.

Activities of the Oil Industry and Their Impacts on Fisheries in the Todos os Santos Bay

From the starting of oil exploration, in the early 1950s, the oil activities in the bay waters affecting fisheries can be summarized as:

- Surveying and drilling of temporary impact;
- Oil and gas extraction in mangrove and bay waters during more than 50 years;
- Oil terminal operations during more than 60 years;
- Decommissioning of the oil field started about 1 year ago (1997);
- Installation of the multiple pipeline on the early 1990s (gas and liquid fuels), with temporary and permanent impacts;
- Installation of the gas pipeline Manati in 2003, with several impacts on sand banks and restrictions to fishing; and
- Restarting the oil extraction in the Dom Joao Mar oil field, from 2007, this plans to build islands of concrete on the mangrove to support diagonal drilling into the sea bed.

Seismic surveys are recognized for their real and potential damages to marine life, either of permanent or temporary character. The process consists of canons firing air pulses which produce compression waves aimed to reflect into the rock layers of sea bottom (MME, 2007). Main impacts are pointed as restrictions to fishing in the study areas, as well as acoustic perturbation of marine life, in particular cetaceous, sirenia (manatees), fishes, turtles, cephalopods, crustaceous and plankton. Damages to these species are reported to be temporary. However, more vulnerable primary rings of the aquatic food chains are reported to suffer permanent damage, as it is the case of eggs, larvae and juveniles (IBAMA, 2003, *cit. by* MME, 2007). The account of these losses on fisheries is yet to be provided by environment impact assessments.

Seismic surveys of the Dom João Mar field were carried out early in 1950-1960, adding to the aggressive chemicals used in the initial stages of oil exploration.

Drilling of the oil field involve a number of operations able to interfere in the characteristics of sea floor and sea water, which are both the environment where marine life find conditions for living and reproduction. Recognized impacts of drilling are the wasting of gravel and fluids from the perforation. Only from 2000, a Federal Law has prohibited gravel discharging in surfaces under 60 metres deep, which includes the Todos os Santos Bay.

International experts acknowledge that drilling mud as one of the most dangerous wastes for marine environment. To improve lubricating capacity a variety of oils have been used, including diesel oil, toxic for marine life. The impact of drilling mud on the aquatic ecosystem is proportional to the volume used in drilling, estimated on 1,000 to 5,000 m^3 per well (Patin, 1999). For the Dom João Mar oil field, local information suggests a number of wells between 500-600, in a sea depth below 10 metres. An aggravating condition is the low renovation capacity of water in the area, as its located at the far end of the bay, opposite to the open ocean. As an estuarine environment, the area is covered by mangroves, where traditionally nest a rich variety of coastal life, especially crustaceous and molluscs, which used to make up an important part of local income and dietary needs of local communities. Both classes of marine life are recognized as highly sensitive to temporary bioaccumulation of petrol and organic components (Pedrozo, M., 2002: 85-86).

Only from the 1980s, the industry started using a less toxic composition of drilling mud, incorporating vegetal oils, less aggressive to marine life (Patin, 1999).

Another risk for the ecosystem has been the water wasted during the oil extraction process. To keep the pressure of the well and force the oil or gas out, sea water is injected into the oil/gas wells. The amount of water injected is increased as much as the oil level reduces, and becomes an indicator of the maturity of the oil field.

"The expression 'production water' designates a chemical effluent made by formation water, injection water and chemical additives. The more mature becomes an oil field, more water will be produced, since the water flow will have to be progressively raised, keeping the reservoir pressure." (MME, 2007: 566)

An study commissioned by Petrobras on the impacts of offshore drilling in depths 10 times of those of Todos os Santos Bay, and in open sea conditions, found alterations on sediments and on Benthic community of the sampled area (Petrobras, 2003, *cit.* by MME, 2007).

Recent analyses of sediments from mangroves and from the neighbourhood of the Madre de Deus oil terminal, where production and transportation has been active over the last 60 years, found significant levels of contamination from hydrocarbons connected to the oil originated from the Bay fields. The study of 2003 provided evidence of *"the presence of petrol and their compounds in the mangrove ecosystem, characterizing a scenery of pollution by organic compounds associated to the petroleum complex."* (Veiga, 2003).

Fig. 8.2: Madre De Deus: Pipelineway from the Landulpho Alves Refinery to the Madre De Deus Oil and Gás Terminal in the Background, Fishing Boats from the Local Fishers Colony (October, 2007).

Mitigation Proposals to Environmental Damages

From the early 1990s, some large initiatives were carried on to expand and modernize activities of oil transportation from production to the large RELAM refinery located right at the bay shores, a closing distance from the Dom Joao Mar oil field. These contracts provided environmental assess studies to comply with governmental licensing legislation. We argue that, in spite of the acknowledgment of environmental damages, potential and actual, the studies commissioned by Petrobras have offered no adequate compensation to the significant losses imposed to fishing communities in the short and the long run.

This situation is probably connected to the low level of bargain power developed by fishers since industry established in the region. A general interpretation for the whole country suggests that fishermen have failed to reach a level of organization able to successfully negotiate with the petroleum industry (Lopes, s.n.t.). As a result, the compensations have always been connected to specific losses from large oil spills, and granted by justice after 10 years fighting (Ministério Publico da Bahia, 2004).

In the case of Todos os Santos bay, mitigation measures proposed by environmental studies are generally intended to bring education programmes to affected communities addressing issues such as:

- Information on Petrobras activities in the region;
- Information on measures taken by the oil industry to control pollution;
- Fishing classes for fishermen, considering the "spawn calendar and local fishing techniques" (Petrobras, 1990: 9.3);
- Health education; and
- Impacts of industrialization on the environment.

These mitigation measures would be conducted by means of "education pamphlet distribution, oral presentations, short courses in schools and churchs, video presentations etc." (Petrobras, 1990: 9.4).

In all cases, mitigation measures provided no compensation to fishermen for their lost income, in temporary or permanent terms, associated to the changes provoked by oil industry on their fishing grounds.

Fig. 8.3: Oil rig left as testimony of the Dom João Mar oil field, near São Francisco do Conde, Bahia. About 600 of these structures were cut and most sunk in the shallow waters of the bay about ten years ago. (October, 2007)

The low bargain power argument, above proposed, is consistent with the fact that early in 2002 the state of Bahia daily newspapers have publicized the condition of the fishermen concerning the Dom João Mar sunk structures, with no visible solution for the situation (Tribuna da Bahia, 11/32002; Tribuna da Bahia, 15/3/2002). Five years later, in 2007, a solitary small floating crane were collecting debris scattered on the fishing zone seabed with no signs of conclusion for the enormous job (Figure 4).

The main consequence of the abandon of the structures for fishing is the damages caused to the nets as the structures remain underwater with no signaling of their presence at all. The losses of fishing nets are not properly compensated, as Petrobras only provide the exact amount of net lost, not accounting for the days fishers are not able to work. This time may amount to five months, according to official documents available in the local fishing colonies.

Oil leakages into the mangroves has also become another consequence of the abandonment of the structures to fishing activities,

Fig. 8.4: The Single Floating Crane responsible for cleaning the Old Oil Structures, with the Crew showing Nets Found Entangled in the Sunk Debris of Dom João Mar (October, 2007)

as the old structures have not been properly sealed or oil remains in the old rusted pipelines (Figure 5).

Relationship Practices between Fishermen and Petrobrás

The central hypothesis in this paper is that exists a direct relation between the attitudes of industry towards local communities and the aggravation of subsistence conditions, leading to an increasing predatory using practices of the estuarine ecosystem. These attitudes from industry reflect a perception of communities as elements alienated from the oil operations, and which can be dealing with social communication strategies informing about the essentiality of services being performed. On the communities perspective, these programmes sound as contradictory and disconnected from the reality facing by families in their working daily life.

In the relationship with fishing communities of the Bay, it is surprising the absence, after more than 60 years of presence of the

Fig. 8.5: Oil Leaking from the Old Abandoned Structures in the Mangrove near São Francisco Do Conde and Oil Sample collected by Local Fishers (October, 2007)

oil industry, of a single permanent dialogue channel allowing the negotiation of environmental damages caused to the local communities. The inexistence, or better the fragility and inconsistency of communications between community and industry seems indicate that meeting production targets will be always more relevant than provoking incidental damages to the ecosystem or the local fishermen. This perspective is underlined by the notion that profits and environmental damages are altogether impacts of oil exploration and shall be expected to increase as an oil field gets matured, as stated by the Brazilian oil industry blue paper:

> "The most important impacts in this phase [when the reservoir pressure tends to fall and becomes necessary to pump gas and water into the well] are the alteration in the water quality and the marine biota due to the wasting of production water, increasing the country's hydrocarbon production and the generation of royalties and special payoffs [to local governments]." (MME, 2007: 566).

EIAs-Environmental Impact Assessment studies reflect a perspective of the national policy level which wrongly put royalties paid by the oil industry to federal and local governments in the same category as compensations for environmental damages. Under this point of view, the industry could use the royalties as a sort of insurance against damages caused to the environment. If that was true, fines imposed by environmental bodies would become redundant. Actually, the use of royalties as a kind of compensation has been part of the discourse of the industry throughout its history. Compensations for environmental damages due to those who depend on biotic resources to make their living have been forgotten in name of the product yielded and the taxes generated by the oil industry. However, the social realities of production areas demonstrate that too little of this resources have actually managed to trickle down to the benefit of sea workers and their communities.

In spite of the technological leadership obtained by the Brazilian oil industry, the production areas of Todos os Santos Bay still resents of social compensatory tools able to regain the output suppressed by, in many cases, inevitable practices of oil exploration activities. Royalties are transferred automatically to the budgets of states and municipalities, to be used in local programmes not necessarily, and in fact not at all connected to the fishermen. No contribution is given by any level of government to fishermen organizations in the whole country. Petrobras resources to social programmes are intermediated by NGOs and spent in short courses to fishermen and women catchers. These activities do not involve direct income transfers to capitalize sea workers.

The technology gap among fishermen may express all the relationship difficulties between industry and fishermen. The sophistication and cost of the infrastructure and logistic used by Petrobras in their exploration activities sharply contrast with the rudimentary canoes — made from a carved tree—and elementary equipments afforded by the artisanal fishers in the region (Petrobrás, 2003). To date, no contribution by the industry to the technological improvement of the fishing sector has been registered.

Conclusions

In general, the Brazilian experience suggests that local communities expect:

- To be recognized as stakeholders deserving attention and respect by government and private sector;
- As a signal of this recognition, be informed, timely and properly, about what, when and where changes are going to happen in the local environment where they live or they earn their subsistence means;
- To be compensated "accordingly" when changes become inevitable;
- That channels be created and maintained open between the agents involved in the process of change;
- That communication at local levels starts before decisions be made or works initiated;
- That communication be held during and specially after the enterprise be concluded; and
- That communication results in concrete advances in terms of life improvements for communities (Barros *et al.*, 2005).

During the past 60 years, the practice of relations between the oil industry and fishermen communities has been aggravated by relationship failures able to be rescued in the medium term. This chapter revises some of the events promoted by the industry in the Todos os Santos Bay, and the impacts as perceived by local fishing communities.

A review of impacts from oil exploration over the last 60 years suggests the urgency of changing corporative attitudes and practices towards fishers, men and women. Constructive relations for widening opportunities to increase income and the opening of new job opportunities suppressed by the oil industry are urgently needed, in face of the degradation of living conditions among fishermen.

Despite the considerable amount of finance addressed by the oil industry to the governments of the region through royalties, the melting of these resources within local budgets prevent fishermen to earn adequate compensations and increase productivity to regain losses caused by oil related activities on their fishing grounds.

NOTES

1. MME. 2007. Plano Decenal de Expansão de Energia 2007/2016, MME/EPE, Brasília, DF.

2. For adequate compensation we understand those resources able to replace the income suppressed by oil activities.
3. Guidelines and Standards for the Removal of Offshore Installations and Structures on the Continental Shelf and in the Exclusive Economic Zone, in IMO WAG. Resolution A.672(16) (1989).
4. Drilling mud is a special mixture of clay, water, or refined oil, and chemical additives pumped downhole through the drill pipe and drill bit. The mud cools the rapidly rotating bit; lubricates the drill pipe as it turns in the well bore; carries rock cuttings to the surface.

REFERENCES

Barros, H., Paranaguá, M., Alves, M.S. Protocolos de Relacionamentos Sociais para Recuperação de áreas impactadas por atividades petrolíferas, Rede Recupetro/Proamb/UFBa, 2005.

Fisheries Research Services. Fishing Gear Impact and Interaction. http://www.frs-scotland.gov.uk/Delivery/Research_monitoring/research_monitoring.aspx, access on 10/2007

Hardin, Garrett. The Tragedy of the Commons, *Science*, 162(1968): 1243-1248.

IBAMA.INSTITUTO BRASILEIRO DO MEIO AMBIENTE E DOS RECURSOS NATURAIS RENOVÁVEIS-DIRETORIA DE LICENCIAMENTO E QUALIDADE AMBIENTAL/ESCRITÓRIO DE LICENCIAMENTO DAS ATIVIDADES DE PETRÓLEO E NUCLEAR (IBAMA/DILIQ/ELPN). Informação ELPN/IBAMA Nº 012/03: Impactos Ambientais da Atividade de Prospecção Sísmica Marítima. Rio de Janeiro, 2003.

Imo Wag. Resolution A.672(16), Adopted on 19 October 1989. Guidelines and Standards for the Removal of Offshore Installations and Structures on the Continental Shelf and in the Exclusive Economic Zone, http://www.imo.org/Safety/mainframe.asp?topic_id=1026, access on 14/01/2008.

Jornal da Tarde. Petrobras é multada pelo Ibama, 2005. *Jornal da Tarde*, Salvador.

9

Innovation, Information System for Economic Infrastructure in Village Planning

H.R. Yadav

1. Introduction

It has been realized that economic development and commendable advances made by the nation in diverse field would in reality be incomplete, if they fail to translate into better and dignified life for the majority of people who live in remote villages. The economic development must be participatory, through tapping the local talent and available economic infrastructure, in order to ensure sustainability of the planned programmes. The planning process for economic development has been implemented aiming at the sustainable development of the village people with a focus on enhancing the livelihoods of weaker and vulnerable sections of the society. It has been the commitment of government to accord foremost priority for integrated village development for which a number of steps have been initiated to formulate planning programmes at grassroots level, i.e. Village Panchayat—Block Panchayat—District Panchayat level and to ensure the implementation of planned ongoing programmes and to make the schemes more effective and accelerated through the people participation and their sustained involvement in the implementation of various rural development programmes, hence ensuring sustainable development of the Village Panchayat—Block Panchayat—District Panchayat.

In this study an attempt has been made to analyze the economic infrastructure including land cultivation, land ownership, irrigation system and area under agricultural production, fertilizer application and other infrastructural facilities which are the backbone for the initiation of policy planning and implementation of need based planning programmes through ensuing peoples participation for sustained integrated development.

2. Objectives

The main objectives of the study is to analyze the agriculture production, land ownership, irrigation system, agricultural implements, animal husbandry and other sources of economy in the village Sarai Khema of the Amethi Block.

3. Economic Infrastructure

The economic infrastructure of an area plays an important role in the formulation of planning process and implementation of integrated village development programmes at grassroots level. Information management for economic infrastructure has been one of the challenging task for the planners and development workers. Various indicators of economic infrastructure needs to be integrated for the use of remotely based rural areas. The collection and collation of economic infrastructure data base in the framework of special information system for optimal management is required for the sustainable development. Proper linkage of economic infrastructure data base within a framework of suitable information system to enable customized retrieval and analysis for specific needs to be managed at village level in Amethi.

Economic infrastructure data base management system is to be visualized in a network based nodes covering villages-block-district. Technological innovations is to be popularized to promote existing abilities to do more things to do the work better and faster. Nevertheless not every innovations that look like productivity—enhancing tool eventually to promote that role. Therefore, if not carefully introduced, it may run counter to natural modes of expectations, thereby leading to rejection of the technology.

Economic infrastructure based local area planning programmes at village level requires routine basis updation of data base to village level, block level and district level, so that as per economic growth, the process of planning can be formulated. The major constraints in the large scale economic data collection and updation of data from grassroots level. It requires technology development and training initiatives for development of appropriate indicators of economic infrastructure data base. It also requires employment of hyper spectral economic data in the application related to agriculture, irrigation, land use, land ownership pattern use of agricultural implements and other related areas of development.

In view of the above, an attempt has been made to explain and innovate the economic infrastructure information system of village Sarai Khema of Amethi Block. It has been assumed that economic infrastructure data base may play an important role in the overall development of the area, therefore, the collection and collation of data base related to economic infrastructure are explained because there are one of the available resources which play an important role in the village development planning and its successful implementation for integrated sustainable development. Thus the various indicators of economic infrastructure data base has been analyzed for the village Sarai Khema of Amethi Block in Chhatrapati Shahuji Maharaj Nagar District because economic infrastructure data base is one of the major important indicators of the natural and human resources information system for the integrated sustainable development planning at grassroots level, i.e. village level.

4. Cultivation

Agricultural cultivation is the main occupation in the village Sarai Khema of Amethi Block. It has been recorded that more than 60 per cent area of the total area of the village is under cultivation followed by the wastelands, old fallow land, fallow land and banjar land. It has been observed that 79 per cent population of the village Sarai Khema is dependent on the agricultural activities which includes cultivators and agricultural labourers. Thus village Sarai Khema is dominated by the agricultural activity. It can be concluded that the agriculture is the major economic activity in the village Sarai Khema of Amethi Block.

There are 207 hectares agricultural/cultivable land in the village. The gross cropped cultivated land of the village is 474 hectares due to multi-cropping including cropping in the Kharif, Rabi and Zaid seasons of cropping. It has been observed that there are variations in the net cultivated area and gross irrigated area due to adoption of multi-cropping pattern which has caused an increase in gross irrigated area as per the fertility status of the land.

5. Land Ownership

Land ownership pattern in the village Sarai Khema is different from other neighbouring area due to kingship, Zamindari, Taluqdari traditions in the village. The large number of farmers are the owners of less than one hectare of land while only 10 per cent land owners are the land owners of more than 90 per cent of the land in village Sarai Khema of Amethi Block. The land concentration in village Sarai Khema is very high because very few land holders are owning more than 90 per cent of the land of the village Sarai Khema, while highest more than 90 per cent land owners are the owner of 10 per cent land of the village.

6. Irrigation

Net irrigated area of the village Sarai Khema is characterized by the mixed cropping. The agriculture is the backbone of economic activity of the village Sarai Khema of Amethi Block. Most of the people of the village are dependent on agriculture. The food grain production is the major agricultural activity in village Sarai Khema, while commercial cropping has very insignificant share in the overall agricultural production of the village Sarai Khema.

Wheat is the most important crop which is cultivated in 40 per cent of the agricultural cropped area of the village Sarai Khema of Amethi Block.

It is estimated that approximately 210 hectares of land is cultivated under wheat crop in Rabi season and similarly it is utilized for paddy cultivation during the Kharif season of the cropping pattern. Approximately 35 per cent area of the total cropped area is cultivated under paddy crop in the village Sarai Khema of Amethi Block.

Barley is cropped in the 5 per cent of the total cropping land in the village Sarai Khema. Millet and Maize crops are cultivated in the very limited agricultural land in village Sarai Khema of Amethi Block.

Gram and pea are cultivated in the limited land. Farmers cultivate gram and pea crop as per their family consumption and requirement. Gram and pea crops are not cultivated as commercial crop in village Sarai Khema of Amethi Blcok.

Mustard, potato and sugar cane crops are also cultivated in the limited land as per their household requirement. The farmers do not cultivate mustard, potato and sugar cane as commercial crop except as per their personal requirements. Sanai and Dhaincha is being cultivated for green manuring and their household needs including to main the fertility status of the soil with the view to increase agricultural productivity of the land in village Sarai Khema of Amethi Block.

7. Agricultural Implements

Agricultural implements play an important role to increase the agricultural productivity of the cultivated land. The use of high-tech and technically developed/upgraded agricultural implements are very nominal because most of the farmers belong to the small and marginal farmers category and their size of land holding is less than one hectare. Due to low size of land holding it has been difficult to utilize the high-tech costly agricultural implements.

Thus application of agricultural implements in village Sarai Khema is very nominal which indicates that the farmers are still utilizing the traditional system of agricultural cropping pattern, which has led to poor agricultural productivity.

Traditional type of wood cultivator is used by the large scale of farmers in village Sarai Khema. Most of the farmers still use iron and wood cultivator for agricultural cropping in the village. Thus more than 90 per cent farmers use wood and iron cultivator which indicates that the farmers are still utilizing the traditional system of agricultural implements which has caused low/nominal agricultural production in the village Sarai Khema of Amethi Block.

The application of high-tech cultivator is comparatively less/ nominal due to small/marginal size of land holding.

Tractor is very nominally used by the farmers. There are only 5 tractors used for cultivation but these tractors are also commercially utilized. Due to small size of land holding application of Tractor cultivation is nominally used by the farmers of village Sarai Khema of Amethi Block.

The application of fertilizers is common exercise among all the farmers of the village Sarai Khema. Nitrogen, Phospherous and Potash (NPK) fertilizers are commonly used by the farmers. Balanced application of NPK fertilizer is not used as per the fertility status of the soil and common minimum requirement of the soil to increase the agricultural production because the requirement of the soil varies from one field to the another field. The ignorance of the farmers in soil testing is not utilized and the farmers have intension to increase the agricultural productivity through increased application of NPK fertilizer in their respective cropping land. The excess application of fertilizer has caused deterioration of the fertility status of soil, henceforth leading to decrease in the agricultural productivity of the land. Thus, scientific and balanced application of fertilizer as per requirement of the soil plays an important role in increasing the agricultural productivity of the farmers in village Sarai Khema of Amethi Block.

8. Animal Husbandry

The most dominant animal in the village Sarai Khema are—cow, buffalow, sheep, goat, horse, pig, hen etc. The cow is one of the important animal in the village Sarai Khema of Amethi Block. The cow is available significantly among most of the families in the village Sarai Khema in the form of worshipping mother cow and also for the milking purposes of female cow and use of male calf for agricultural cultivation including other household activities.

The buffalows are also significantly distributed among most of the families in village Sarai Khema of Amethi Block specially for the milking purpose.

The distribution of sheeps and goats are significantly distributed in the village Sarai Khema of Amethi Block.

The pigs and hens are important animals among the scheduled caste and Muslim families respectively in the village Sarai Khema of Amethi Block. Most of the Kori/Chamar households are having the pigs while most of the Muslim households are having hen/cock for their families purpose especially chicken, eggs etc. use the commercial application of pigs/hen etc. are very insignificant except personal use among the most of the families.

Thus the animal husbandry activity is carried out mostly as per the family requirement while their use for commercial purpose is very insignificant due to small and marginal farmers and low size of land holding in village Sarai Khema of Amethi Block.

9. Source of Economy

It is known fact that village Sarai Khema of Amethi Block has been one of the backward villages of the District Chhatrapati Shahuji Maharaj Nagar. Most of the people of the village Sarai Khema are socially economically backward. Most of the villagers survive under below poverty line, whose income per annum is less than Rs. 60,000.

The common average income of the farmers of the village varies between Rs. 9,000 and Rs. 1,00,000 per year. The average income of the farm labourers and common labourers varies between Rs. 36,000 and Rs. 50,000 per year which is very nominal due to which such people are forced to survive hand to mouth.

The skilled labourers of the village are able to generate an income upto Rs. 96,000 per annum, while secondary and tertiary workers economy varies as per their educational/technical qualifications including the business workers. High income group persons are very nominal which is very insignificant in comparison to low income group population. Thus it is evident that most of the 95 per cent population of the village Sarai Khema are surviving under below poverty line and most of them belongs to low income group.

10. Conclusion

It is known fact that economic infrastructure of the village Sarai Khema plays an important role in the formulation of integrated planning process and implementation of village development. The collection and collation of economic infrastructure data base in the framework of spatial information system for optimal management of the sustainable development.

Cultivation has been the main occupation in the village Sarai Khema because more than 60 per cent area of the village is used under cultivation, and 79 per cent of population of the village is dependent on the agricultural activity. Thus the agriculture is main source of income in the village Sarai Khema.

It is evident that 90 per cent households are the owners of 10 per cent land of the village while only upper caste 10 per cent households are the owners of 90 per cent land of the village.

More than 300 hectares of the land is irrigated land. Canal, government tube well, and private tube well are the main source of irrigation in village Sarai Khema.

The agriculture is the backbone of the economic activities in the village Sarai Khema. Wheat, paddy, barley, pea, gram, potato, sugarcane etc. are the major crops being cultivated in the village Sarai Khema.

Traditional agricultural implements are still used in the village Sarai Khema due to small size of land holdings. High techno-agricultural implements are not used due to ignorance and socio-economic backwardness.

NPK fertilizer and insecticides and pesticides are used by all the farmers to increase agricultural productivity without agri-scientific and technological backup support in the agricultural production.

The cows, buffalows, pigs, hen, cock and other need based animal husbandry activities are carried by the small/marginal/large farmers including landless agricultural labourers as per their household needs.

Most of the villagers survive under below poverty line except very nominal households of the village Sarai Khema. Thus it can be concluded that village Sarai Khema is not only socially backward but most of the people of the village are economically backward and these people are forced to survive employment generating and income increasing programmes has to be planned and implemented with the people.

10

Sustainable Yogic Farming: A Social View

U. Yadav and *Shailendra K. Singh*

"Everything else can wait but not agriculture."

Agriculture is the way of living in India and it is the mother of all cultures and that is why it is called "No culture without Agriculture". Because of agricultural experts and farmers, India has been able to produce 242 million tonnes of foodgrains in 2011-12 against only 50 MT in 1950. We are self-sufficient in foodgrains but still lag behind in pulses and oilseeds. India is a country of villages and farmers specially 91 per cent small and marginal farmers better known as peasants and the prosperity of this great nation passes through the farming and fields of these peasants. These marginal farmers particularly women farmers with an area below one hectare are 77 per cent in numbers in India and backbone of Indian Economy and they are the blood and soul of Indian progress and prosperity.

In fact, India is rich country kept poor by selfishness, greed, pride and ignorance of her people, especially of those educated and big farmers who occupy positions of authority and power in various walks of life, viz. political, religious, economic and social and those who control the economic and natural resources of India and that is why more than 30 per cent, i.e. about 40 crores of Indian population are still unable to have one good nourishing meal rather half fed. More than 20 million children in India, everyday go to sleep in the night getting tired of crying for food. More than Rs. 65,000 crores of worth of foodgrains are rotting away and/or eaten by rats and termites in the FCI Godowns of India. More than Rs. 35,000 crores worth foodgrains are stolen from PDS and sold in the black market every year. About 80 per cent farmers want to leave farming business in case of any alternative solution is available.

Today India is facing a great challenge of degraded natural resources like soil, water, plants, air and above all human being. Because of intensive agriculture practices, farmers are using heavy indiscriminate doses of chemical fertilizers and pesticides out of ignorance and greed. This results in soil erosion, soil acidification, loss of soil microbial activities and great increase in pollution level leading to floods, famines, droughts etc., poor soil health and fertility.

In the light of these facts and figures available to us, one can easily conclude that the real problems facing India today are not economic and political but spiritual and moral agriculture, better known as 'sustainable agriculture'. Hence, the lasting solutions to these problems can be found only through a spiritual sustainable farming resolution of unprecedented order. So, innovative concept like Sustainable Yogic Farming through "5M" by Indiahs and world farming community specially small and marginal farmers involve application of (1) Microbial Organisms, i.e. Organic manures (Green manuring FYM, vermi and NADEP compost etc.), Bio-fertilizers, (2) along with practice of Meditation, i.e. Positive Thought Energy, (3) Music, (4) Mantras, and (5) Agricultural operations as per Movement of Moon. Researches have proved that all these said practice, in combination can transform Indian agriculture in respect of change in DNA, higher percentage of seed germination, increase in quality grain, healthier root, stem and leaves growth, increase in grain's self-life and right kind of production and productivity. We will discuss it one by one in the following ways:

(1) Microbial Organisms (Organic Farming)

Organic agriculture is the only answer for sustainable agriculture growth. The sustainable agriculture may be defined as "Any set of agronomic and organic practices that are economically viable, environmentally safe and socially acceptable. It recommends the use of such agricultural practices which are biological in nature and eco-friendly. These agricultural practices preserve environment and biodiversity. Organic agriculture is a holitistic production management system which promote and enhances agro-ecosystem health including biodiversity, biological cycles and soil biological activity. It emphasizes the **use of the management practices in preference** to the **use of off-farm inputs** taking into account that regional conditions require locally adopted system.

- Organic farming experiments in rice based cropping (including wheat and vegetables) system through microbial inputs have shown higher yield at IARI, New Delhi. Pusa Basmati 1121 grown organically with four inputs, viz. Blue Green Algae, Azolla, Vermi-compost and FYM gave the highest grain yield, i.e. 4.46 tonne/ha which was follows by the yields of crop grown with Integrated Nutrient Management (INM) (4.32 T/ha) and Chemical Fertilization (4.18 T/ha).
- To minimize the post-harvest losses, bio-agents like *bacillus subtilis, Pseudomonas flurosence, D hausenii* etc. could be effectively used for the control of post-harvest diseases in strawberry and kiwi fruits.
- Application of bio-fertilizers, viz. Azotobacter + PSB in newly released varieties of wheat (DBW 17 and PBW 550) was popularized in Faridabad (Haryana). Treatment of these varieties with **bio-fertilizers** resulted in an increase in yield up to 6.5-7.3 per cent over that of the control.
- Use of new herbicide molecules for weed control in wheat was popularized using METSULFURAN @ 20 kg/ha in Haryana which resulted in 51.40 per cent yield increase over that of hand weeding and no use of any chemical.
- Use of zero tillage for wheat cultivation practice in Gurgaon district of Haryana during 2008 and 2009, resulted in 5.06 per cent increase in yield over that of control plot apart from saving on cost of production up to Rs. 4000-5000 per hectare.
- Bed planting is a technique in which crops are sown in a raised bed with a bed planter and irrigation applied in furrows. Bed planting helps to increase productivity of Kharif season crops like maize, pigeon pea, green gram, soybean which are subject to waterlogging and water deficit. Beds formed in Kharif season can also be used without any preparatory tillage for sowing of succeeding crops.

Better agriculture practices should focus on the following areas to achieve the goal of sustainability in agriculture:

(i) Reduced use of off farm inputs with less harm to environment and consumers. Bio-fertilizer and bio-pesticides are the potential alternatives in the area.

(ii) More productive use of biological and genetic potential of plants and animals. Biotechnological tools can help us to create novel technologies which have higher potential.

(iii) Better match between cropping patterns and the physical capacity of lands with better agricultural practices and harnessing the natural resources.

(iv) Improved emphasis on conservation of soil, water energy and biological resources as the excessive use of chemical inputs has resulted in degradation of our soil ecosystem.

(v) Better water management like sprinkler and drip irrigation which reduces the wastage of water. Rainwater harvesting should be important component of better agricultural practices to increase water availabilities and recharge our ground water sources.

(vi) More thorough incorporation of natural process because natural processes always enrich our environment and biodiversity.

Organic farming helps in creating an ecological balance and a micro-environment suitable for health and growth of soil micro-flora, plants, animals, farm workers and finally, the vast population of consumers.

Organic farming is practiced in more than 150 countries and 35 million hectares of agricultural lands are managed organically. In India about 5,28,171 hectares area is under organic farming with 44,926 number of certified organic farms. This account for about 0.3 per cent of total agricultural land. Indian organic farming industry is estimated at US 78 million dollars and is almost entirely export oriented.

The resource conservation practices conceive for this include minimum of zero tillage, letting crop residues get back into the soil instead of burning them, immaculate land levelling to ensure the even spread of water and applying only need-based fertilizers and water to crops. They protect soil health to enhance its fertility, prevent the environment pollution caused by burning of crop residues, save on the labour and energy required for repeated land tilling, and reduce the use of water in agriculture, sparing it for other purposes. The biggest advantage is that by letting biological residues get back into the soil, it transforms agriculture from carbon emitter to virtual carbon requester by converting crop land into a carbon sink. Organic agriculture can be understood in the following one line told by renowned Agriculture Poet GHAGH:

"जाको खेत में पड़ा न गोबर, उस किसान को जानो दूबर"

Lord Krishna in *Gita* has advocated organic (sustainable) farming in chapter 18 (44) as . "कृषि गौरक्ष्यवाणिज्यम्", i.e. Agriculture along with animal husbandry and business is the ultimate solution for sustainable agriculture growth.

Women farmers in India are an asset to Indian agriculture and they contribute almost 95 per cent agricultural operations starting from sowing of seed till marketing of produce but they have not been given due importance as they deserve. So women farmers must be given top priority in organic agriculture because they perform agricultural practices very skilfully, emotionally and with perfection.

(2) Meditation I.E. Positive Thought Energy

The acceptance and recognition of thought power is increasing worldwide. One of the important reason in increasing this is the awareness of non-monetory inputs for safe farming and food as well as powerful tool to improve health of all lives. The positive thoughts have been found to change DNA, water, seed germination growth and development of crop plants as well as health of people. It is a powerful tool to create above said changes even from distance.

Scientific research has proved that all vegetations are lively and have sensitivity (J.C. Bose). Vibrations of human beings and sensitivity of nature definitely affect the crops and plants. Agricultural produce too, would therefore be deeply influenced by the positively charged vibrations of Rajyoga Meditation to yield nutritious crops in greater margins at a lower cost. Successful experimentation through Rajyoga has already been proved.

Sustainable Yogic Farming, an innovation concept involves practice of Rajyoga Meditation along with the application of organic manures and biofertilizers and successful results were obtained from trials conducted at S.D. Agricultural University, Dantiwada, Gujarat under an MoU for 5 years with Brahma Kumari Spiritual University, Mt. Abu, Rajasthan. On the basis of review regarding effect of Rajyoga Meditation, i.e. Positive Thought Energy on plants. Scientists of G.B. Pant University of Agriculture and Technology, Pant Nagar are motivated to conduct field trials on some of the crops. Important findings of various researches are (1) improved percentage of seed germination, marginal increase in test weight, healthier root modulation, enhanced soil micro-biological activities, higher content in groundnut, drop off in the pest-damage, increase in nutritional values, enhanced phosphorus content in the soil, appreciable increase in appearance and taste of vegetables and increases shelf life. There are some of the experiment results which have been found as follows:

- Barrington, *et al.* (2003) found that there was greater germination rate and growth of seedlings in the groups that were watered with energized water that had been treated with positive thought of the healer.
- Effects of qi (healing process) on the germination rate of rice seeds was that percentage of seeds that germinated was generally greater for qi-treated seeds than that of the control. This was because of the vibration of positive thoughts created by healer with positive power.
- Leikam (1981), Nash (1982), Elizabeth (1983) showed that distant healing with power of thoughts produced significant effects on enhancing and retarding growth of bacteria and yeasts.
- Rein and Me Craty (1994, 1955) showed that the implications of these studies are far reaching indeed that the distant healing with positive thoughts could alter the rates of winding and unwinding of strands of DNA.
- In the field of Datatric Akram Nale in Sangrual Village, he got higher price of Gur produced from sugarcane under meditation plot compared to chemical fertilizer plots.

Thus the power of thoughts has been found to play a vital role in transforming agriculture as well as human health. The ill effects of modern farming will be minimized and safe and secured food production will be achieved and the people/farming population will be more healthy and wealthy. The power of Rajyoga, i.e. Positive Thought Process can make it success as it is a nomonetary powerful input for farming system.

Not only this, villagers constructed bunds to check soil and water erosion and conserve moisture *in-situ* and they constructed ponds to provide life saving irrigation for plants and life saving water for animals in summer. Now-a-days, this whole village is flourished with several crops in a year and people are very happy and prosperous.

Almost every Indian is well aware of the power of Mantras which can bring desired change (positive or negative) in every walk of life and Mantras are now-a-days being recognized as the healer against disease and pests and the food products are qualitative and quantitative both, giving a good and right income to farmers. Mantras are the group of positive sounds made up of combination of different words. These sounds give positive effect on senses of animals and plants while negative sounds deplete energy of both plants and animals. Mantras as NADA and it is known as Brahmanada have impact on mind and body. Regular observation of Yagna, immunity of plants increased and thus more production with less insect, diseases pest is achieved.

(3) Music and Agriculture

Music can create wonders and that is why it is called "सुर साधे, सब सधै" meaning thereby music can accomplish everything. Whole universe is musical and you have simply to realize and utilize it for the betterment of humanity and agriculture. How can we forget the flute of Lord Krishna, when he used to play the flute, not only human beings but animals, birds and even snakes and plants get attracted in a very positive manner and get energized for better results and all worries of the world are banished. Researches have proved that music on the agriculture field not only enhances quality, right production and productivity but grains are very healthy and growth is very faster. In our traditional agriculture, music was being produced through the bells tied in the neck of bullocks and cows which induced positive energy in animals, farmers and farming resulting into more production in low cost. Women farmers and labourers sing SOHAR song while transplanting paddy uprooted from nursery and song vibration helped roots to establish earlier than the plants without song. Singing while planting, weeding and harvesting has been our ancient tradition giving very very good results in respect of production and soil fertility.

It is believed that India is the origin place of music and if plants are given music for half an hour daily, results faster growth and higher productivity of plants.

(4) Movement of Moon and Crop Yield

Crops sown as per movement moon gives higher production with no costs. The moon affects earth's ocean tides and currents which causes weather and climate patterns. The moon moves regularly around earth and it travels monthly through earth of the 12 signs of the zodiac, staying approximately two and half days in each sign. As it does so it forms an angular relationship with the sun and that is known as a phase of the moon, which means the angle between Moon, Earth and Sun. Moon orbits the Earth and Earth revolves around the Sun. it is the Earth's orbit that defines the Ecliptic which is divided symbolically into Zodiac.

Lunar month is divided into two basic moon activities, i.e. waxing and waning. The beginning of each process is designated as New Moon and Full Moon. Full moon occurs when moon is at 180° opposite to the Sun, in position to receive sunlight over the maximum volume of its surface. A new moon occurs when the sun and moon are so closely aligned that is impossible for moon to give off any reflected sunlight.

The phases of moon occur in two stages, i.e. Waxing and Waning moon. (1) The Waxing moon is waxing in ascending period between the New Moon and Full Moon. Plants are growing these phases and it is good time to encourage plant growth and proliferation. As a thumb rule when moon is waxing plants develop leaves above ground system and planting leafy crops grow above ground are best sown. (2) Waning moon happens in descending period. Waning describes lunar activity between the full moon and new moon. Plant starts shrinking above the ground system. As a thumb rule, when moon is Waning plants develop their root system. Waning moon phase is a useful time to control plant growth and it keeps down pests.

Planting leafy crops that grow above ground are best sown at Waxing Moon and those that will require strong root system or grow belowground should be sown after Full Moon, in the Waning phase.

Perigee (*Poornima*: Full Moon) when the moon is nearest to the earth, this occurs after 29 and half day. In 48 hours, preceding to Full Moon, there appears to be distinct increase in the moisture content of the Earth. Growth promoting activities of the plants seems to be enhanced and plants were susceptible to fungal attack because of relatively higher moisture content in rhizosphere and atmosphere.

Apogee (New Moon) when the moon is farthest from the earth. This occurs every 27th and ½ day. Owing to moisture deficiency, harvesting and seed storage practices show better response.

Moon opposite to Saturn is a favourable period, agricultural operations performed device during this period show better response.

Lunar Node on imaginary point when moon crosses path of Sun. It occurs twice in 27.2 days of a month and known as Rahu and Ketu.

Rahu—Lunar node in ascending period of "Moon" is not suitable for agricultural activities.

Ketu—Lunar node in descending period, is not suitable for agricultural activities.

In order to harness the cosmic forces, agricultural operations including bio-dynamic preparations should be prepared and utilized as per bio-dynamic agriculture calendar.

These kinds of technology, even if not very practical in current scenario because of lacking trained healers, experts and mental make up of policy makers and authorities, must be promoted for future or new era likely to come very soon. This will not only transform agriculture but also facilitate transforming farming families and other human beings.

(5) Mantra/Words Power in Transforming Agriculture

"God made nothing lovelier than words." Words are most solid and powerful tool to translate ideas into action. Mantras have got unique power as Words/Mantras are Brahma (spirit). Gandhiji gave a simple slogan "English Quit India" and whole Nation, under one word, forced English to quit India and that is the magic of words. Subhas Chandra Bose said "Give me blood, I will give you freedom." All Indians were behind Bose. Very recently Anna Hazare called upon the Indians for i.e. India against corruption and whole nation, young, children, adults and old were on the road behind this slogan and this is the importance of words.

Ashok Srivastava, Pradhan of Village Uchchar in Rae Bareli District of U.P., used the inspiring lines of the poem "Karmavira" by great poet Ayodhya Singh Hariaudh in order to energize the villagers to transform the agriculture. The inspiring lines are:

"देखकर बाधा विविध बहुविध घबराते नहीं।
सैकड़ों मरुभूमि में नदियाँ बहा देते हैं वे।"

Almost all villagers were inspired and did a uphill task of digging a drain of 3.5 km. long and 1.5 metre deep.

Future of India lies in the happy blend of sustainable farming and spirituality. So in order to ensure food security to all Indians and eternal (perpetual) agriculture, Sustainable Yogic Farming (SYF) through '5 M', i.e. Microbial Organism; Meditation, Mantras, Music and Moon movement by peasants specially women farmers (larger production by larger hands) must be encouraged and given top priority by planners, policy makers, agricultural scientists and big farmers in the forthcoming Twelfth Five Year Plan in the larger interest of the nation.

"कृषि मूलश्च जीवनम्"।

REFERENCES

A New Era through Shaswat Yogic Krishi—Prajapita Brahma Kumari Iswariya Vishwa Vidyalaya, Mt. Abu, Rajasthan.

A Review on Harvesting Thought Energy be fitting for transformation of Agriculture: A Paradigm Shift—Sunita T. Pandey, Omvati Verma, Kewalanand and D.S. Pandey, Deptt. of Agronomy, College of Agriculture, G.B.P. University of Agriculture and Technology, Pant Nagar, Uttarakhand.

Better Practices for Sustainable Agriculture Health and Environment, *Kurukshetra Journal*—Dr. Harender Raj Gautam, Dr. M.L. Bhardwaj, Dr. Y.S. Pawar, University of Horticulture and Forestry, Naini, Solan, H.P.

Bio-dynamic Agriculture—R.K. Pathak and RA. Ram, Central Institute for Subtropical Horticulture, Rehmankhera, Lucknow.

Dainik Jagran.

Effect of Lunar on Crop Yield.

Farm Technology for Enhanced Productivity and Income.

Hindustan Times.

How the Moon affects Weather and Climate.

Jaivik Kheti—Aaj Ki Jarurat—Yadunath Singh Murari.

Krishi Gyan Manjusha.

Role of Science and Spirituality in Sustainable Agriculture, IARI, New Delhi.

Second Struggle of Freedom.

Times of India.

Visit Report for Brahmakumari Institute, Kolhapur, Maharashtra by the Director of Research, S.D. Agriculture University, Sardar Kaushi Nagar, Maharashtra.

Yojana, October 2011.

11

Innovation, Information System for Institutional Infrastructure for Village Development Planning

H.R. Yadav

1. Introduction

Availability of institutional infrastructure established by the Government, Semi-Government, Panchayati Raj Institutions and civil society institutions plays an important role in the overall development of an area. The available infrastructure of any area plays a key role in the formulation of planning process. The process of integrated approach between various sectors of economy to resolve often conflicting requirement and available infrastructure of the area. It leads to a requirement for appropriate data management specially the sectoral data related to institutional infrastructure. Therefore, it becomes necessary to analyze the institutional infrastructure in the selected villages of Amethi. As per availability of the institutional resources the common minimum requirements of the down-trodden people can be realized and need based people planning process can be initiated. Plan without peoples participation and without due consideration of the common minimum requirements infrastructure people participatory planning process cannot be made successful. Therefore, available institutional infrastructure must be planning process at grassroot, level.

In view of above, an attempt has been made in this study to analyze the available historical, religious, administrative, communication, drinking water, energy, medical health, agro-industries, health services, educational infrastructure etc. to ensure the need base planning process with the involvement of the local people and to ensure need based people's participatory need base planning and its successful implementation with the people, by the people and for the people for the sustainable development in village Sarai Khema of Amethi Block. Thus it need a proper planning for the growth and development of the socio-economic institutional infrastructure of the village Sarai Khema.

2. Objectives

The main objectives of the study is to analyze the administrative, irrigation, drinking water, communication, educational, industrial, medical health, holding and other infrastructure available in the village Sarai Khema of Amethi Block.

3. Historical and Religious Infrastructure

Historically and religiously village Sarai Khema has insignificant importance. The village has been under the rule of Rajputs, Hindus and British due to which the size of land holding is concentrated within 10 per cent population of the village. Historically and religiously, there are three old temples which are still maintained and the local village people used to worship in the temple. Thus the religious infrastructure of the Village Sarai Khema is very important and locally famous. Religious temples located in the village Sarai Khema may be promoted for religious tourist centres with the view to generate employment among the local people.

4. Administrative Infrastructure

Village Panchayat Sarai Khema of Amethi Block comprises three revenue villages, viz. Dadupur, Sarai Khema and Ram Nagar Mafi, while these three revenue villages are having more sub-villages. Thus revenue village Dadupur comprises Dadupur Khas, Pure Kutur and Jagapur sub-revenue villages. Sarai Khema revenue village comprises Sarai Khema Khas, Munshiganj, Basuapur, Halapur (big), Beni Madhav Kutia and Ahmad Nagar sub-revenue villages and Ram Nagar Mafi revenue village comprises Purepandit and Mainapur sub-revenue villages of village Panchayat Sarai Khema of Amethi Block.

5. Irrigation Infrastructure

Canal and tube wells are main source of irrigation in the village. Sarai Khema of Amethi Bock. The total length of canal within the village Sarai Khema is more than 2 km. There are two government tube wells as source of irrigation. There are 12 electric engine operated tube wells in the village Sarai Khema of Amethi Block. There are 60 private tube wells owned by the individual farmers for the irrigation of their own cultivated fields. Thus private onward tube wells are one of the major source of irrigation in village Sarai Khema of Amethi Block.

6. Drinking Water

Traditionally, wells have been the major source of drinking water in the village Sarai Khema. Presently, there are more than 21 wells in the village, which are rarely used for the purpose for which they have been constructed by the village people. There are 28 private hand pumps installed by the village people individually, while approximately 290 India Mark II Hand pumps are installed by the various Government agencies from the different development funds by the District Administration in village Sarai Khema of Aemthi Block. Thus most of the India Mark II hand pumps has been installed by the government agencies, including fund from the Member of Legislative Assembly (MLA), Member of Parliament Local Area Development (MPLAD) and implementing agency Jal Nigam at local level. Thus the drinking water infrastructure of the village Sarai Khema has been installed as per the needs of the local village people.

7. Health Infrastructure

Primary health services has been one of the most important basic requirement of the local village people. There is sub-health centre in the village and one A.N.M. Centre is also functional in the village Sarai Khema with the view of the private common minimum health support services at grassroots level in the village. To ensure health and minimum nutritional food and educational support services at grassroots level there are 5 Anganbadi Centres in the village Sarai Khema of the Amethi Block in the Amethi district of Uttar Pradesh.

8. Village Panchayat

There is one Panchayat Bhawan constructed in the village Sarai Khema of Amethi Block for official use of the village Pradhan to control the educational, revenue, health, common and child development, rural sports and strengthening the Panchayati Raj Institutions in village Sarai Khema of the Amethi Block.

9. Rural Roads

Road infrastructure plays an important role not only facilitating the communication network but interaction support system with the development programmes implementing agencies. The village Sarai Khema is adjacent to Hindustan Aeronautics Ltd. (HAL), Korba, Munshiganj and Sanjay Gandhi Hospital and marketing centre Munshiganj which is well connected with the Amethi Block and Tehsil headquarter which is only 4 km. away from Amethi Town Area. Average distance of *Tarkol* constructed road is approximate more than 3 km. from the village Sarai Khema of the Amethi Block. There are 3 km. *kharanja* road constructed within the village Sarai Khema and its revenue villages and sub-villages while more than 2 km. *chackroad* has also been constructed in the village. Thus is evident that village Sarai Khema is well connected with the rural road network facilities.

10. Communication

There is no sub-post office in the village, villagers use to go to Munshiganj and Amethi for postal facilities. There are 3 public call centres, and telephone public support facilities are available in Munshiganj and Amethi adjacent to the village Sarai Khema of Amethi Block. It has been recorded that more than 80 per cent people of the village Sarai Khema are the user of the mobile phone facilities. Thus the telecommunication facilities and awareness for its use at grass- roots level makes it clear that the village Sarai Khema is socially educationally and economically developed and award to certain extent.

11. Educational Infrastructure

The educational infrastructure of the village Sarai Khema is well developed starting from primary education to the higher and technical education due to the HAL factory, hospital and Block and Tehsil headquarters at Amethi which is well connected by the road networks within the village Sarai Khema of the Amethi Block of the Amethi District.

There are 3 primary schools in the village Sarai Khema, one Junior High School and two Intermediate Colleges. There is one Balika (Girls) Inter College in the village Sarai Khema. It has also been noted that there is one Pharmacy College, and one Technological and Management College, Technical, Education and Higher Education Centre in the village Sarai Khema of the Amethi Block,

12. Banking

There is no Banking facility centre in the village Sarai Khema while at the distance of approximately 2 km Bank of Baroda and Uttar Pradesh Baroda Gramin Bank are functioning in the Munshiganj, which is a hamlet of village Sarai Khema including State Bank of India functioning in the HAL (Korwa) campus.

13. Agro-Industries

It has been recorded that there is no significant agro-industrial development in the village Sarai Khema of the Amethi Block. There are 3 flour mills (Chakki) in the village and 2 rice mills (chakki) in the village Sarai Khema. In spite of this there are two oil mill, in the village Sarai Khema. Thus, it is evident that the flour, rice and oil mills are established by the individuals to fulfil the common minimum needs of the village people, which has not only facilitated the villagers but it has also generated employment and increased the income of the involved local people of the village Sarai Khema.

14. Small Scale Industries

There is very insignificant development of the small scale industries in the village Sarai Khema of the Amethi Block. It has been recorded that there are 2 Poultry Farms (hen) in the village Sarai Khema and one Biscuit Bakery in the village is also functional. There are 55 commercial shops operational within the village Sarai Khema of Amethi Block. Thus it can be concluded that need based small scale industries and small level commercial shops have been opened due to the Amethi Town Area and H.A.L. factory, hospital and other marketing centres especially in Munshiganj which is adjacent to the village Sarai Khema of Amethi Block. It has been observed that such commercial shops and small scale industries functional in the village Sarai Khema are not helpful at larger scale to generate employment and increase the income of deprived rural multitudes of the village.

15. Energy

It has been observed that most of the houses of village Sarai Khema are electrified and maximum village streets are also electrified. Electricity facilities have been provided among the down-trodden section of the society under Rajiv Gandhi Gramin Vidyutikaran Yojana and most of the households have also taken electricity connection in their respective households.

Commercial consumption of the electricity and commercial electricity connection is very insignificant except the commercial shops, small scale industries and agro-industries established within the village Sarai Khema of the Amethi Block. The electricity has been one of the main sources of energy in the village Sarai Khema.

16. Animal Hospital

As per the policy of the government, there is one sub-animal hospital, functional temporarily nearby the Panchayat Bhawan Sarai Khema. One Animal Doctor and a staff occasionally used to be present for providing animal hospital facilities to the needy village people. Thus the role of animal hospital and its facilities are very insignificant for promotion and upgradation of animal in the village Sarai Khema of the Amethi Block.

17. Housing

Low cost housing schemes known as Indira Awas Yojana has been facilitated to the Below Poverty Line (BPL) people. Housing schemes are implemented by the Pradhan, Village Panchayat Sarai Khema, local area people's representatives funds and District Rural Development Agency (DRDA), District Chhatrapati Shahuji Maharaj Nagar. It has been recorded that more than 55 households belonging to below poverty line (BPL) and landless had been given housing facilities. It has been observed that low cost housing facilities are not properly constructed as per the norms of the government and needs of the beneficiaries has been ignored, due to which people's participation in the construction of such rural houses has been found very insignificant.

Thus the housing is one of the basic needs of the deprived down trodden people living below the poverty line and the shalter/housing facilities are the basic needs of the man kind in terms of safety, security, social status, cultural identity, self-satisfaction and achievement due to which the housing for all is the goal of the government. Keeping the above facts into consideration rural housing facilities has been supported in the village Sarai Khema for the needy resource less deprived in multitudes and the criteria laid for the selection of beneficiaries for Indira Awas Yojana (Rural Housing) which has been the primary need of the people.

18. Conclusion

Thus the available institutional infrastructure of an area plays an important role in the integrated development especially for the formulation of need based planning process and implementation of the peoples participatory planned programmes. Plan without peoples participation and without due consideration of the needs of the people and available institutional infrastructure and people's participatory planning process cannot be successfully implemented. Therefore, available institutional infrastructure has to be considered before initiation and formulation of planning process at grassroots level.

There are three historical religious temples in the village Sarai Khema of Amethi Block, Village Panchayat Sarai Khema comprises three revenue village, viz. Sarai Khema Khas, Dadupur and Ram Nagar Mafi along with 12 hamlet villages, Munshiganj is one of the hamlet villages which is marketing centre and having facilities like Hospital and other higher educational centres.

Canal and Tube wells including Government Tube wells and Private Tube wells are the main source of irrigational infrastructure in the village Sarai Khema of the Amethi Block. There are 290 India Mark II hand pumps and 28 private hand pumps installed for drinking water facilities. Health infrastructure of the village Sarai Khema is available as per the needs of the local people. Village Panchayat

Bhawan has been constructed in the village Sarai Khema to provide administrative support system to the Village Pradhan and members of the village Panchayat Samiti.

Rural road infrastructure has been constructed as per the needs of the local village people. Banking facilities are available in one of the hamlet villages at Munshiganj in village Sarai Khema.

Agro-industries and small scale industries are operational as per the needs of the local village people. Most of the villages of the village Sarai Khema are electrified and commercial electric connection has also been provided for small scale industries and commercial shops opened in the revenue and hamlet villages. Animal Hospital is also operational in the village Sarai Khema. Below poverty line resourceless need householders have been provided rural housing, i.e. low cost houses because the housing is one of the basic needs of the common people.

Thus institutional infrastructure of village Sarai Khema has been developed as per needs of the local people which has to be strengthened with the view to generate employment and increase the income of the deprived rural multitudes of the village.

12

'E-Krishi Kiran'—Village Information System for Scientific Farming

A.M. Shekh and *Arun Patel*

e-Krishi Kiran programme is expected to bridge the gap between Scientist-Extensionist-Farmers and Input-Output Dealers effectively. It helps making transfer of technology more scientific, precise, easy, and need based. It is a web based information system designed to run on a networked environment including intranet and internet. This is a repository of agricultural information for the benefit of farmers, agricultural scientists and policy makers.

Government of Gujarat has selected the Anand Agricultural University (AAU) to establish and implement the state-wide programme for *e*-extension. Under this programme, AAU has developed the web-base application software, which generates and provides the fertilizer recommendations on the basis of soil analysis and the nutrient requirements of the crop for each field. This will increase the efficiency of the fertilizer and saves consumption of the fertilizer. The home page of the web base application of e-Krishi Kiran Programme (Gujarati language) is given as **Appendix-A.**

Fine Objectives

- Analysis of soil of all the villages of the state.
- Recommendation for fertilizer uses for each farm, based on soil analysis.
- Suggest alternate cropping pattern based on moisture availability index.

- Provide Soil Health Card to each farmer to guide him about his soil conditions and cropping practices to be followed.

Features of the SHC Application Software

- Gujarati, Web based application, designed to run on network environment, intranet, Internet and GSWAN.
- Secured access to the planners, scientists and other users based upon unique identification.
- User friendly navigation.
- Auto generated built-in alarms and messages.
- Uniformity of user interfaces for easy understanding and operation.
- Generation of various kinds of qualitative, quantitative and analytical reports.

Structure of the e-Krishi Kiran Programme

(1) Input

This part includes input of Agricultural Data like—(1) Farmers record with their Name, District, Village, Survey number, SHC number, etc., (2) Soil fertility data based on soil test report regarding organic carbon, available phosphorus and potash, EC and *p*H, (3) Soil type and depth, (4) Talukawise rainfall data, (5) Crop cultivation practices, (6) Recommended NPK, (7) Alternative crop planning, (8) Crop economics data, (9) Action plan information (information of Taluka/Village/individual action plan):

- The provision has been made for entry and modification of the input data in the main server at Gandhinagar and mirror server at AAU, Anand by respective institutes/officers sitting at their places/offices (state Govt. Deptt./District/Taluka/State Agril. University etc.).
- Data sets of various modules of Soil Health Card are highly secured and can be accessed only through valid username and password for respective entry and modification.

(2) Database and Software

Following provisions are made in the software system:

The System can Process

- For the validation of the soil analysis data, so that data entry can be checked by the system and wrongly entered data may be rejected during the process of entry of data.
- Various transaction data for soil, weather, fertilizers, profitability of a crop, etc. for generation of various reports and recommendations for the crop, so that desired reports and recommendations can be made available.

The System can Generate

- Recommendation of the fertilizers needed for a particular crop based on a nutrition status of farmer's soil on the basis of low, medium, high soil fertility rating.
- The soil fertility index of each village on the basis of available soil analysis reports of the insisting farmers, to generate village soil health card for the use of other farmers who had not made soil testing of their fields.
- Recommendation of the possible alternative crops to a farmer for better crop production based on his cropping practice and weather condition in his area considering the facts like moisture availability index, available water capacity, length of growing period, surplus water and supplementary irrigation.
- Recommendation of the alternative crops with a generic ranking of crops in terms of profitability, looking into the various costs involved for producing that crop.
- The statewise, districtwise, talukawise and individual farmerwise model action plans for crop production.
- The various FAQs for farmers about agriculture and animal husbandry in a well classified manner.

(3) Network Connectivity

Network for the e-Krishi Kiran Programme is the part of GSWAN (Gujarat State Wide Area Network) with main server at Gandhinagar and its mirror server at Anand Agricultural University, Anand; connected with the three sibling agricultural universities (NAU, JAU, SDAU) of Gujarat; development departments of Gujarat (agriculture, horticulture, animal husbandry); GAIC, GSSC, GLDC, GSSWDC, GSSCA, districts offices, taluka offices, KVKs, etc. The system can also run with any type of internet connectivity.

(4) Output

(A) Soil Health Card

Soil health and its fertility play a key role in crop production. Soil analysis is the means to know the soil health. However, till recent past, fertilizer recommendations for different crops are made on the basis of agronomical practices and not on the soil test base.

Soil test based recommendation for fertilizer use will not only increase the crop production with judicious investment on fertilizer use but will also help to keep the soil productivity sustainable.
The English version of printed farmer's soil health card is given as **Appendix-B.**

The information given in Soil health card can be classified in four main groups as per details mentioned hereunder:

(a) **General Information:** (1) Soil health card number, (2) farmer's name, (3) name of the village/taluka/district.
(b) **Land Information:** (1) Account number, (2) Survey number, (3) Survey number-wise land area, (4) Soil type.
(c) **Soil analysis information:** Value and explanation of result for (1) *p*H value, (2) Electric conductivity, (3) Organic carbon, (4) Available phosphorus, (5) Available potash.
(d) **Crop-wise fertilizer recommendations:** (1) Season, (2) Crop name, (3) general recommendations of dose for chemical fertilizers (N, P, K) made for each crop by the Department of Agriculture, (4) Recommendations of dose for chemical fertilizer (N, P, K) made for each crop on the basis of soil analysis report of each field, (5) Recommendation of FYM for each crop.
(e) **Ready Reckoner for calculating the quantity of fertilizer:** On the backside of the village soil health card and farmer's soil health card ready-reckoner for calculating the quantity of fertilizer is given so that farmer can easily calculate the quantity of Urea/DAP or other chemical fertilizer to supply prescribed dose of N P K.

(B) Village Soil Health Card

In general it is perceived that the land of different fields in a same village may have less variation in respect of the soil fertility. Hence, the fertility index is calculated from the available reports of the soil analysis of the fields of same village and on the basis of fertility index, the farmers who had not made their soil tested were recommended for the requirement of the fertilization of each crop. Most of the other informations remain same as per the individual soil health card discussed above.

Access to Soil Health Card

The individual soil health card or the village soil health card can be viewed/accessed through internet. There are two options in the systems to view the soil health card information. First option is to enter soil health card number directly. If the farmer does know his soil health card number, then he has the second option of selecting his district, taluka, village and farmers name. Listing of soil health card numbers of the selected farmer is displayed which are hyperlinked. Once a particular hyperlinked soil health card number is clicked, the soil health card is displayed.

However, it is presumed that the farmers may have not easy access to the internet hence Government of Gujarat has made provision to supply printed copy of the individual soil health card to each farmer. Moreover, the printed colour copy of the village soil health card will also be supplied to each gram panchayat during Krushi Mahotsav regularly every year.

(C) Alternative Crop Planning

In soil health card programme the crop planning for rain fed crops is based on LGP (length of growing period). The LGP is derived from MAI (moisture availability index), which in turn is obtained from AWC (available soil moisture capacity). In this exercise, soil type, bulk density, available water in soil, rainfall (average of 80 years), PEI etc. are taken into consideration. Surplus rain water available for recycling through farm pond is also obtained. These data have been worked out for all the talukas of Gujarat.

Farmers require the guidance on the best possible alternative crop which can be grown on his farm considering soil and meteorological parameters. Moisture availability index could be the base for the suggestion of crop planning.

The availability of water in the soil during the course of crop growth determines largely its yield. Soil water availability depends on rainfall, PET, types of soil and crop water use. Though the availability in amount and distribution of rainfall and PET (across location) undoubtedly affect the duration and characteristics of growing season, the differences in soils, specially in respect of moisture storage and release characteristics strongly influence the agricultural value of rainfall. Thus, a study of soil water balance is a prerequisite in planning rain-fed farming for optimum rainfall use.

So, for each agro-climatic zone, existing crops and cropping systems and suggests new crops based on LGP and cropping systems has been suggested. Economics was also worked out and net profit for each crop/cropping system was derived.

(D) Model Action Plan

A Model Action Plan is implemented at district, taluka and village levels as a part of Soil Health Card Programme. Model Action Plan for each taluka/village can be viewed in detail from the Internet.

For farmers' benefit and effective adoption of the programme, the following information is included in this plan:

(1) Characterization of natural resources such as soil, water, climate, generally grown crops, manpower etc.
(2) Statistics of crop production, market price, etc.
(3) Information on changes in cropping patterns, productivity, pest and diseases, increasing salinity, water logging or rise in ground water table etc. along with reasons.
(4) Planning required for implementation of specific agriculture or village development programme.
(5) Types of extension programme required.
(6) Achievement of goals during the year and how the programme can be modified.
(7) Government and non-government organizations (NGOs) working at each taluka level.
(8) An integrated work plan of village.

(E) FAQ and Solution of Queries

The Question-Answer bank is maintained in a classified way for the frequently asked questions by the farmers. The answers of such questions are prepared by the concerned scientist, so that first hand knowledge of research recommendations may be available to the farmers.

If a farmer requires more details of technology or has any problem he can send his problem through the query form available on the system. The query form is designed in such a way that farmer has to made the selection of the text, name of district/taluka/village, etc. and very few words are required to type. Further, arrangement has been made in the application that the query will automatically reach to the concerned scientist/extension expert for immediate action through the system. Besides, a printed copy of the answer will also be posted by the scientist/experts. Thus, it would not be necessary for the farmer to visit Internet, but the answer to his query would reach him by post itself.

Appendix-A

The Home Page of Soil Health Card Programme (Gujarati Language)

Appendix-B

English version of Farmer's Soil Health Card

Paper on: 'e-Krishi Kiran' - Village Information System for Scientific Farming

A.M. Shekh and *Arun Patel*

e-Krishi Kiran programme is expected to bridge the gap between Scientist-Extensionist-Farmers and Input-Output Dealers effectively. It helps making transfer of technology more scientific, precise, easy, and need based. It is a web based information system designed to run on a networked environment including intranet and internet. This is a repository of agricultural information for the benefit of farmers, agricultural scientists and policy makers.

Government of Gujarat has selected the Anand Agricultural University (AAU) to establish and implement the state-wide programme for *e*-extension. Under this programme, AAU has developed the web-base application software, which generates and provides the fertilizer recommendations on the basis of soil analysis and the nutrient requirements of the crop for each field. This will increase the efficiency of the fertilizer and saves consumption of the fertilizer. The home page of the web base application of e-Krishi Kiran Programme (Gujarati language) is given as **Appendix-A.**

Fine Objectives

- Analysis of soil of all the villages of the state.
- Recommendation for fertilizer usages for each farm, based on soil analysis.
- Suggest alternate cropping pattern based on moisture availability index.
- Provide Soil Health Card to each farmer to guide him about his soil conditions and cropping practices to be followed.

Features of the SHC Application Software

- Gujarati, Web Based application, designed to run on network environment, intranet, Internet and GSWAN.
- Secured access to the Planners, Scientists and other users based upon unique identification.

SOIL HEALTH CARD

DEPARTMENT OF AGRICULTURE
GUJARAT STATE

Year : 2008

SOIL HEALTH CARD NO.:SHC00027842

(1) Farmer's Name : Pursotambhai Hirabhai Patel

(2) Village : Zarolla **Taluka :** Borsad **District :** Anand

(3) Account No : 280

(4) Land / Soil detail as per account :

Serial	Survey No.	Area (Ha.)	Soil Type
1	1004-2	00 : 40 : 00	Sandy loam
2			
3			
4			
5			

(5) Fertility class as per soil analysis of the village :

Serial	Fertility Class	Nitrogen	Phosphorus	Potash
1	Low	✓		
2	Medium		✓	
3	High			✓

(6) Details of individual soil analysis : Survey No. : 1004-2

Sr.	Detail	Result	Interpretation of Result
1	pH (Soil Reaction)	7.20	Normal
2	E.C. (Total Dissolve Salts: dSm/m)	0.49	Normal
3	Organic Carbon (%)	0.26	Low
4	Available Phosphorus (kg/ha)	43.00	Medium
5	Available Potash (kg/ha)	254.00	Medium

pH		E.C. (Electric Conductivity)	
Acidity :	6.5 or less	Normal :	Less than 1.0
Normal :	6.5 - 8.2	Medium :	1.0 to 3.0
Alkaline :	8.2 or high	Harmful :	3.0 or high

(7) Crop wise Fertilizer Recommendation on the basis of soil analysis.
(Supply recommended fertilizers as basal dose and in top dressing)

Note: Soil characteristics illustrated in this card is pertaining to this survey number only. Soil characteristics vary for different survey numbers. Use of fertilizers in accordance with soil analysis of individual field is more beneficial.

Season / Crop	General recommendation of fertilizers by Dept. of Agriculture (kg/ha) N	P_2O_5	K_2O	FYM (Tonnes/ha)	Recommendation on the basis of soil analysis (kg/ha) N (kg/ha)	P_2O_5 (kg/ha)	K_2O (kg/ha)
Kharif — Banana					(*FYM-kg/plant, NPK-gm/plant)		
Banana*	180	90	180	15	200	90	150
Kharif — Paddy							
Paddy (Drilled)	80	20	0	6	90	20	0
Medium late (T.P.)	50	25	0	10	55	25	0
Early Maturing (T.P.)	100	25	0	10	110	25	0
Late Maturing (T.P.)	120	30	0	10	130	30	0
Kharif — Tobacco							
Bidi Tobacco (Hybrid)	220	0	0	15	240	0	0
Bidi Tobacco (Improved)	180	0	0	15	200	0	0
Kharif — Castor							
Irri. Castor	75	50	0	10	85	50	0
Kharif — Pearl millet							
Hy. Pearl millet (Rain fed)	80	40	0	15	90	40	0
Kharif — Chilly							
Chilly	100	50	50	10	110	50	50
Rabi — Culcutti Tobacco							
Guj. Culcutti-1	200	0	0	10	240	0	0
Guj. Culcutti-2	150	0	0	10	165	0	0
Rabi — Wheat							
Late sowing	80	40	0	15	90	40	0
Timely sowing	120	60	0	15	130	60	0
Rabi — Chickpea							
Chickpea	20	40	0	10	25	40	0
Rabi — Potato							
Potato	200	100	200	25	220	100	200
Summer — Paddy							
Summer Paddy	100	30	0	10	110	30	0

- User friendly navigation.
- Auto generated built-in alarms and messages.
- Uniformity of user interfaces for easy understanding and operation.
- Generation of various kinds of qualitative, quantitative and analytical reports.

Section III
Rural Development Programmes

Introduction

Rural Development in our country is an important segment of integrated development because most of the people live in Rural India. An integrated rural development has been felt and commendable advances have been made in the diversified field of development would in reality be incomplete, if such Rural development programmes fail to translate into better and dignified life for the most of the people living in the remote rural area. Rajesh Kumar Nayak has tried to explain the Nation and Rural Development in India and images of Gandhi and Nehru.

H.R. Yadav has made an effort to analyze the implementation of varioius rural development programmes being implemented at village level in Amethi. In this study an attempt has been made to analyze the various rural development programmes including MNREGA, Energy Small Scale Industries, Sanitation, Rural housing welfare programmes, Agriculture, Social-forestry, Fisheries, Rural Development and the Micro-Credit and financial support and implementation of various rural development programmes being implemented at village level are explained in this study.

Jones Lewis Arthur and Michael Kofi Nsiah have made an attempt to analyse the contemporary approaches to sustainable development exploring critical innovations in traditional leadership in Ghana.

H.R. Yadav has tried to explain the role of Master of Social Work Students in empowering deprived rural women for sustainable development. Self Help Groups of the women beneficiaries have been formed and banking support has been provided to the women Self Help Groups (SHGs) beneficiaries. Practical training and demonstration of Preservation and Processing of fruits and vegetables, its packaging and marketing has been ensured to generate employment and increase the income of deprived rural poor women beneficiaries for their self-sustenance has been analyzed in this study.

Vivek Kumar Singh has tried to analyze the development of India through integrated rural development programmes.

H.R. Yadav has attempted to explain the need based village

development planning at village level in Amethi, District Amethi, U.P. Need based Village Development Planning for the important amenities, viz. health, sanitation, education, drinking water, food, welfare programmes and other schemes being implemented and supported by the government has been analyzed and need based planning process has been formulated to fulfil the ever increasing demand of the local village people.

Mahendra Kuamr Padhy has analyzed the communication strategy for sustainable development through the Indian experiment, various implications of the sustainable development have been discussed in this study.

Ravi Suryavanshi and Neha Nigam have attempted to explain the role of community Radio in rural development based on communication. Today community radio is playing a significant role in rural development because community radio serves people living together and sharing same interests in the different geographical areas. The radio service is able to appreciate and understand specific needs and aspirations of the local people etc. have been analyzed in this study.

13

Nation and Rural Development in India: Images of Gandhi and Nehru

Rajesh Kumar Nayak

Rural Development in India is an important segment of economic development because the real India is rural India. It is a concept, a phenomenon, a strategy and a discipline. The term is of focal interest and widely acclaimed in both developed and developing countries. As a concept, it connotes overall development of rural areas, that is, villages, with a view to improving the quality of life of rural people. It is a comprehensive and multidimensional concept and encompasses the development of agriculture and allied activities, socio-economic, infrastructure, community services and human resources in rural areas.

Stalwarts of Modern Indian history like Mahatma Gandhi and Jawaharlal Nehru often pointed out to the colonial constructs of Indian society in the nationalist's imaginations of India. The rural area, that is, the village was an important category where such communities could be easily observed. For them, the village represented India in microcosm, that is to say in words of Gandhi, "*India begins and ends in* villages."[1] For India as a nation, village was the important unit which transformed into a large entity. Apart from its methodological value, it being a representative unit of Indian Nation had been an important ideological category in modern Indian imagination. The village was not merely a place where people lived; it had a design in which were reflected the basic values of Indian civilization.[2] In this context it was obvious that the development of village, that is, rural development was a hard core reality for the Indian Nation's development.

Focusing primarily on writings of Gandhi and Nehru, the paper attempts the interrelation of village and nation, the ideological concepts of Gandhi and Nehru for rural development and its relevance in modern context.

Any colonial country that is trying to attain its independence, or just have achieved it, is faced with a number of problems arising not only from social unrest but also from an underdeveloped political and economic structure moulded primarily to serve imperialist needs. Gandhi and Nehru, both among the most outstanding leaders of the colonial India developed highly intelligent and pragmatic rather than abstract visions for the solution of these fundamental social, political, and economic problems shadowing India. They understood the crucial importance of grassroots level development, as an estimated 80 per cent out of 400 million Indians inhabited the over half a million villages scattered around the country, most of them not even accessible by proper roads.[3] Furthermore, the number of fully employed people was at about 15 million, in addition to the 25 million employed for only a few months in a year.[4]

Gandhi and Nehru worked very closely during their over a quarter of a century-lasting relationship, and despite their often conflicting opinions and a sharp difference of approach to many of prevailing problems, they exchanged ideas in an atmosphere of mutual respect, and thus greatly influenced each other.

Gandhi's exaltation of the village and village communities is well known, as is his view that cities were places of evil and corruption. Nehru, on the other hand, thought of the village as a place of backwardness and ignorance. In a letter to Gandhi in 1945, he wrote, "I do not understand why a village should embody truth and non-violence. A village, normally speaking, is backward intellectually and culturally and no progress can be made from a backward environment".[5] Not only was the village a place of backwardness and ignorance, but British rule had also robbed it of its organ city and vitality; it had come "progressively a derelict area, just a collection of mud huts and odd individuals".[6] Yet, Nehru added, the village held something of value, "But still the village holds together by some invisible link and old memories revive". Indeed, despite his distaste for the village and for the hierarchical social structure it represented. Nehru's 'The Discovery of India' also contains a powerful sense that the countryside was an authentic symbol of India. Thus, recalling the effect of Gandhi on the Congress and nationalist mobilization, he wrote, "He (Gandhi) sent us to the villages, and the country side hummed with the activity of innumerable messengers of the new gospel of action. The peasant was shaken up and he began to emerge from his quiescent shell. The effect on us was different but equally far-reaching, for we saw, for the first time as it were, the villager in the intimacy of his mud-hut..... We learnt our Indian economics more from these visits than from books and discourses. The emotional experience we had already undergone was emphasized and confirmed."[7] The emotional resonance that both Gandhi and Nehru found in the village can be understood in terms of Partha Chatterjee's argument that "the nationalists identified the 'inner' as the nation's authentic space."[8] The village stood for the domain where the nation was sovereign, free from the 'outer' sphere of modern politics, economics, science, and technology that was dominated by the West. What Chatterjee does not recognize sufficiently, however, is that the logic of the 'inner' was overwritten with that of the 'outer'. The very specification of the authentic and sovereign domain of the nation was also an act of outlining strategies of transforming the 'Inner' with the knowledge and power produced in the 'outer'. Gandhi's conception of the village community was the real India. It is clear from the appendix of 'Hind Swaraj', which mentions among other 'authorities', Henry Maine's text on village communities, that Gandhi drew on European writings in formulating his notion that India resided in its villages. This, in itself, is not significant. But it assumes importance when considered in conjunction with his conception of the village. In 1945, Gandhi outlined his vision of the village in a letter to Nehru, "While I admire modern science, I find that it is the old looked at in the true light of modern science which should be reclothed and refashioned aright. You must not imagine that I am envisaging our village life as it is today. The village of my dreams is still in my mind. After all, every man lives in the world of his dreams; My Ideal village will contain intelligent human beings. They will not live in dirt and darkness as animals. Men and women will be free to hold their own against anyone in the world. There will be neither plague, nor cholera nor smallpox; no one will be idle, no one will wallow in luxury.... It is possible to envisage railways, post and telegraph offices etc. For me it is material to obtain the real article and the rest will fit into the picture afterwards."[9] Gandhi was candid about the idealized nature of his village, but from what position does he idealize it? How should we understand the dream of a village where people did not live "in dirt and darkness as animals?" Evidently, Gandhi was trying to envision an India different from that of Nehru, but it is clear that the imagined village was also a product of the 'dream work' of modernity. The act of 're-clothing' and 're-fashioning' was to apply to both the old and the new, to both the 'inner' and the 'outer', to both the village and the city. The failure to recognize this double movement of the discourse is to overlook the effects of the mutual imbrications of the 'inner' and the 'outer', and conclude that nationalist politics represented the triumph of one over the other. Such a reading is unable, to appreciate the complex moves entailed in the projection of the nation as a modern village, and produces the widely held but mistaken belief that Nehru appropriated Gandhi's concept of the village as the nation's 'inner' sphere and built the nation-state in the image of the 'outer'. Elsewhere, despite significant differences, both Gandhi and Nehru shared the discourse of a 'different modernity', that the establishment of the nation-state cannot be understood as the victory of one domain over another but as a historical development fashioned on the site of the intersection of the 'inner' and 'outer'.[10]

Gandhi "Bapu" (father) as Nehru also called him, indeed, was the person who had drawn Nehru into the midst of the Indian politics and the nationalist struggle, and into real contact with the "naked hungry mass" or rural India. Gandhi acknowledged that his heir had a different vision of the direction of India's development than he had, but this did not change the special connection he felt to Nehru. Though believing in the words of Hind Swaraj until the very end.[11] Gandhi truly wanted to understand Nehru's visions of the development of India, and didn't try to convert Nehru from working towards the direction that he considered right.[12] However, towards the end of his life Nehru clearly started to put an increasing emphasis on Gandhi's teachings, by becoming critical towards the all-India large-scale projects, and speaking for the necessity of linking the 'scientific approach with the spiritualist approach'.[13] After all, Nehru didn't consider the true success of planned economy to be the achievement of certain ambitious statistical targets, but a rise in the standard of living of the common people. In his speech at the Community Projects Conference in May 1952 Nehru stated: ... you can safely say that 96 per cent of the people of India are economically very backward. Indeed, apart from a handful of men, most of the people are backward... we have to think more of those who are backward because we must aim at progressively producing a measure of equality in opportunity and other things."[14] Later on, in September 1958, he wrote, "Somehow we have got entangled in mighty or expensive schemes, which no doubt are good in themselves, but which take a long time to yield results... All this leads me to the conclusion, that the basis of all our work must be intimate contact with our people in almost every phase of our activity."[15] In his letters to the Chief Ministers of the provinces Nehru proposed investment in smaller schemes requiring less capital but showing quicker results, and recommended assistance to be given to the village panchayats to have them engaged in the execution of the plans.[16] In May 1963 Nehru recalled the real, still unachieved, objectives of the country: "...a life worthwhile for all the hundreds of millions who live here so that all of us can lead a purposeful existence and rid ourselves of the curses of poverty, unemployment, disease and ignorance".[17] Nehru's main concern in early 1964, while his own health was getting worse, was how to improve the Third Five-Year Plan so that to make some real improvements to the life of the rural poor.[18]

Gandhi's influence on Nehru, though, should not be underestimated, as it was he who drew young Nehru into real contact with the poor of India, and devoted his whole life to this mission of improving the lives of the underprivileged. However, only in his older age, after Gandhi had already passed away, did Nehru start to realize that he didn't have the miraculous powers of transforming India from a backward society to an industrialized state, and that the basis of all development was at the very grassroots level. The major contradiction between Gandhi and Nehru in regard to their plans on the economic development of the villages arouse from their basic difference of opinion on the way the problems of India could be solved. Nehru wanted the villages to be effectively integrated into the industrial system of India, adopting its methods and equipment, and utilizing India's resources according to their best capacity, as this was the way he envisaged India could rid itself of its prevailing problems. For Gandhi the villages represented, or he wanted them to represent, all the values he advocated and which he thought were the cornerstone of development: non-violence, asceticism, high degree of morality, and the value of each living creature as an essential part of the civilization in the light of this viewpoint, Gandhi's and Nehru's actions and plans in regard to the economic development of the villages can be better understood, as, although they shared many common objectives, they were rarely of the same mind on the ways in which to achieve those goals. However, due to their close relationship, which had deepened in the course of the nationalist struggle, they were open to each other's ideas to an extent that didn't force them to deviate too extensively from their own fundamental principles. Gandhi's ideas on the economic development of the villages were widely criticized by the advocates of modernization. Yet the failures of modern economic planning to rid India of the extreme poverty and the consequent problems might indicate that this was perhaps not the most successful path of development. Despite the assiduous development of India through five-year plans and through the adoption of the Western methods and equipment, 70 per cent of the inhabitants of India still live in villages, and 30 per cent of the people live below the poverty line. Perhaps Nehru would have needed to give a greater emphasis to the economic development of the villages right from the beginning, and thereby to build the basic foundation for further progress, rather than trying to unsuccessfully ruse into the industrial world right away.[19]

Thus, it is possible to look forward to a development scenario in India over the next few decades in which Gandhi-Nehru's perspective on rural development will have great relevance. Nehru and Gandhi were visionary who helped India in retaining the sense of direction in development planning through many vicissitudes and crises. The Indian society is now entering a phase where their perspective and sense of priorities in rural development would become more urgently relevant than before. There would be tough and challenging years ahead in our progress. Achieving Gandhi and Nehru's perspective is likely to be painfully slow. This should be taken as an invitation to redouble our efforts rather than abandon the perspective itself as irrelevant.

In spite of constitutional directives, Legislative enactments executive decisions in the form of various programmes and schemes, the basic problem of rural development has remained almost untouched so for. The search for alternative approaches scheme, administrative innovations and experiments are still going on. The right scheme and strategy is still a far cry. The feudal character of local administrators, representatives and the burden of casteism and individualism are creating an impediment in rural development. The basic defect of out planning process lies not only in our planning but in our thinking for the planning. To solve the problem of rural developments the bridge between village and nation can only be fulfilled if we think about the rural need of rural people that is according to the Gandhi to wipe out every tears of every Indian. This can be achieved through inclusive development of the rural people, that is to say, the development for the people, of the people, by the people and most importantly with the people, which will definitely justify Gandhian and Nehruvian image of rural development.

REFERENCES

1. Gandhi, M.K., *A letter sent to Nehru*, written on August 23, 1945.
2. Beleille, C.A., *Local Roots of Indian Politics*, Delhi, 1980, p. 108.
3. Paul, Jordon, *Modern Times A History of World from 1920 to the year 2000*, London, 2000, p. 469.
4. Frank, Mores, *Jawahar Lal Neharu: A Biography*, Macmillan, 1957, p. 431.
5. Nehru's Replay to Gandhi, 9 October in M.K. Gandhi: *Hind Swaraj and other writings*, Anthony J. Parel ed. Cambridge 1997, p. 152.
6. Nehru, J.L., *Discovery of India*, 1948, Delhi, p. 523.
7. *Ibid.*, pp. 361-62.
8. Chatterjee, Partha, *The Nation and Its Fragments*, Princeton, 1993.
9. Gandhi to Nehru, 5 October, 1945 in M.K. Gandhi, *Hind Swaraj and other writings*, pp. 150-51.
10. Prakash, Gyan, *Another Reason: Science and Imagination of Modern India*, Princeton, 1999, p. 6.
11. David, Arnold, *Gandhi*, Pearson, 2001, p. 20.
12. Brown, Judhith M., *Gandhi: Prisoner of Hope*, London, 1989, p. 365.
13. Nanda, B.R., *Jawahar Lal Nehru: Rebel and Statesman*, Delhi 1995, p. 200 *vide* cited by Nanda in *Statesman* published on 1 December 1958.
14. Frank, Mores, *Jawahar Lal Nehru: A Biography*, New York, 1957, p. 434.
15. Nanda, B.R., *Ibid.*, p. 200.
16. *Ibid.*, p. 216.
17. *Ibid.*
18. Ylitalo, Essi, *The Role of Village Industries in the Development of India as Envisioned by Mahatma Gandhi and Jawahar Lal Nehru*, and Article, Tokyo, 2003.
19. *Ibid.*

14

Innovation, Information System for Rural Development Programmes

H.R. Yadav

1. Introduction

An integrated rural development has been felt and commendable advances have been made in the diversified field of development would be in reality incomplete, if such rural development programmes fail to translate into better and dignified life for most of the people who are living in the remote rural areas.

The implementation of rural development programmes can be made more effective through people's participation. Thus the rural development should be for the rural people and it should be mintained and sustained through the people. Participatory planning and development should be initiated at grassroots level. It has been mentioned in the construction of India that, "Government should evolve methods to unite the village Panchayat and provide such powers and rights, which can be helpful to work as unit of self-Government."

In view of the above, as per policy of Government of India, the State Government of Uttar Pradesh established Self Government of rural areas. Historical and revolutionary decisions have been taken to allot the work, to allocate financial reserves, to implement the planned programmes and also allocated financial and administrative rights to Block Panchayat and District Panchayat.

In this study an effort has been made to analyze the various rural development programme including MNREGA, Energy, Small Scale Industries, Sanitation, Rural Housing, Welfare Programmes, agriculture, social-forestry, fisheries, rural development and the financial support allocated to village pradhan, Block Pramukh, Zila Panchayat, MLA/MLC, MP (Lok Sabha and Rajya Sabha) and Civil Society institutions.

The rural development programmes are being implemented in the village Sarai Khema, in which people's participation are not properly ensured and improper implementation of sanctioned programmes and maintenance of the assets of the implemented programmes has led to the least sustainability of the implemented/ completed programmes because successful implementation of rural development programmes will generate sustained employment and increase the income of the deprived rural multitudes.

2. Objectives

The main objectives of the study is to analyze the various rural development programmes being implemented in the village Sarai Khema of the Amethi Block.

3. Rural Development Programmes

It has been realized that rural development programmes should be participatory and it should be planned according to the needs of the local deprived multitudes. The rural development should be planned through tapping the local talent and available infrastructure of the area in order to make the implemented developmental programmes self-sustainable various rural development programmes are being implemented aiming of sustainable development of rural areas, with a focus on enhancing the livelihoods of weaker and vulnerable sections of the society. An attempt has been made at various levels to make the implementation of on going rural development programmes more effective and accelerated. Development unit shall be village in the new system of decentralization, village shall be the central local point for the development. In reality, it will be meaningful to transfer this in the planning and development programmes, its control and management will be ensured through peoples participation. Planning process will be ensured from the bottom (grassroots level) to the top level. Revolutionary steps of decentralization will bring total change in the life of rural villagers and it will provide a new era for an integrated development of rural areas.

The real meaning of democracy is the importance of the people and rural self-government is the main sole of the rural development including financial management, administrative control of the village Panchayat, peoples participation and co-operation to establish village Panchayat for self-government, for people's participatory planning and implementation of rural development programmes for self-sustenance.

Allocated works of village Panchayat and village Pradhan being the head/chairperson of rural development programmes are mainly education, Tubewells, hand pumps, youth welfare, Medical and Health, Women and child development food supply, agriculture and rural development, Panchayati Raj Institutions, pension and scholarships etc.

Village Government

↓

Village Panchayat

↓

Allocated Works

↓

1. Education—Primary and Secondary Education.
2. Government Tube wells.
3. Hand Pumps.
4. Youth Welfare.
5. Medical and Health.
6. Women and Child Development Programmes.
7. Animal Husbandary/Hospital.
8. Food supply/Ration Shop.
9. Agriculture Department.
10. Rural Development.
11. Panchayat Raj Institutions.
12. All types of Pensions and Scholarship.

In this study an attempt has been made to describe various rural development programmes being implemented in the village Sarai Khema of Amethi Block. Analysis of various rural development programmes include programmes supported by the Government of India, State Government, District and Block Panchayats and Fund from the village Pradhan and funds utilized by the various people's representatives in village Sarai Khema of the Amethi Block in the District Amethi.

(i) **Mahatma Gandhi National Rural Employment Guarantee Act (MNREGA)** is one of the employment generating and income increasing programme in the village Sarai Khema of the Amethi Block. In the MNREGA need based plans are formulated and implemented by the Job Card holders who belongs to the below poverty line. MNREGA created durable community assets, socio-economic infrastructure in the village Sarai Khema. This programme is self-targetting in nature with special emphasis to provide wage employment to the scheduled caste, below poverty line households, and women beneficiaries. There are 498 job card holders in village. Village Pradhan and Panchayat Secretary supervised the work for construction of an ideal pond in the village and Kharanza Chakroad has also been constructed under the MNREGA programme which has not only created assets in the village but it has generated employment and increased the income of deprived rural multitudes of the village Sarai Khema of the Amethi Block.

(ii) **Pradhan Mantri Gram Sadak Yojana** is another important scheme with the view to provide road network, generate employment and better access for agricultural transportation, education, other public services and in general better improvement in the quality of life of the people and bridging the rural-urban divide.

(iii) **Rural Housing** is the basic needs of the mankind in terms of safety, security, self-respect, social status, cultural, identity, satisfaction and achievement. Housing for all is the goal of Government. To meet the shortage of rural housing in the villages "Indira Gandhi Avas Yojana" has been launched. Under the Indira Awas Yojana 55 low cost rural houses have been constructed in the village Sarai Khema with the view to provide shelter to the needy resourceless down-trodden section of the society.

(iv) **Land Reform Schemes** are being implemented and all the revenue records including Khatauni and Khasara registers are updated and computerized. Computerization of Land Revenue Records (CLR) and strengthening of Revenue

Administration and updating of Land Records (SRI and ULR) has been implemented to facilitation the land holders and farmers. The consolidating of the land is also being implemented by the revenue officials to facilitate the farmers for increasing agricultural productivity of the village.

(v) **Rural Water Supply Programme** has been implemented for providing drinking water facilities to the down-trodden section of the society because water is an essential ingredient of our daily life. The importance of safe drinking water cannot be underestimated. Poor quality of water causes many diseases. In view of the above 290 India Mark-II Hand Pumps have been installed in the village Sarai Khema with the view to provide good quality drinking water facilities to the needy villagers.

(vi) **Central Rural Sanitation Programme** has been implemented to remove the disposal of human excreta of cesspool, open ditches, pit latrines, bucket system etc. It is well known fact that direct relationship exists between water, sanitation and health, consumption of unsafe drinking water, improper disposal of human excreta and lack of personal and food hygiene have been the major causes of many diseases. High infant mortality rate (IMR) is also attributed to poor sanitation. As per policy of the government 35 rural sanitation/latrines have been constructed for the scheduled caste families and the people surviving under below poverty line in the village Sarai Khema of the Amethi Block.

(vii) **Energy** has been backbone of any development programme. Most of the villagers have been given Electricity facility while complete electrification facilities have been provided to the scheduled caste households including street light facilities under the Rajiv Gandhi Grameen Vidyutikaran Yojana. Non-Conventional energy development has also been promoted in which solar lights has also been installed in the important places of the village Sarai Khema of Amethi Block. Commercial Use of electricity and non-conventional energy sources, viz. solar lights etc. have also been installed at Munshiganj Hospital and other individuals as per their needs.

(viii) **Small Scale Industries** have been installed in the village Sarai Khema. There are 2 hen/cock farming centres and Biscuit Bakery in the village Sarai Khema of Amethi Block.

Besides this, there are 55 commercial shops at Munshiganj and also in the village Sarai Khema to fulfil the marketing needs of the village people.

(ix) Agro-Industries have also been established in the village Sarai Khema. As per needs of the local people 3 Flour Mills, 2 Rice Mills (Chakki) and 2 Oil Mills have been installed in the village Sarai Khema of the Amethi Block. There is also food supply shops to provide grain for the Below Poverty Line (BPL) Card Holders and Antyoday Yojana. There are 138 BPL Card Holders and 94 Antyodaya Card Holders in the village Sarai Khema of the Amethi Development Block.

(x) Welfare Programmes are launched in the village Sarai Khema which include Scheduled Caste, Backward Caste, Minorities Welfare Programmes. Various welfare programmes, viz. scholarships, financial support for self-employment and other support system for needy interested down-trodden section of the society has been launched. Scholarships are being provided to all the Primary, Juniors High School students including Intermediate and higher education students belonging to the low income group or the parents surviving under below poverty line free-ship and scholarship facilities are provided to the interested needy students belonging to the scheduled caste, backward caste, minority families of the village Sarai Khema of Amethi.

(xi) Rural Road Network has been recognized as simple physical success in the form of an all weather road is itself a powerful catalyst to economic development. Rural road connectivity has become an important instrument in the strategy for alleviating poverty. As a subsidiary objective the programme also seeks to improve the conditions of Rural Roads so as to ensure synergy. It is expected that an expanded and innovated rural road network will lead to agricultural growth, increase in rural employment opportunities, better access to health, education and other public services, general improvement in the quality of life and in bridging the rural-ruban gap. Pucca and Kuccha roads have been constructed in the village Sarai Khema and Kharanja Road and chack roads has been constructed as per needs of the local village people and people's respresentatives in village Sarai Khema of Amethi Block.

(xii) **Agri-culture, Horticulture, Social Forestry Programmes** are launched in the village Sarai Khema. Kisan Mitra Yojana, Kisan Credit Card, National Agriculure, Group Insurance Scheme, Social Forestry and Horticulture programmes are launched in the village Sarai Khema. Very insignificant fisheries programmes have been implemented at individual level in the village Sarai Khema of the Amethi Block.

(xiii) **Medical Health Centres** are functioning in the village Sarai Khema. There are one A.N.M. Centre in village Sarai Khema and 5 Anganbadi Centres are also functioning in the village Sarai Khema. Besides this there is Sanjay Gandhi Hospital at Munshiganj in the village Sarai Khema of Block. There is temporary animal hospital centre in the village Sarai Khema. There are 195 medical health card holders in the village Sarai Khema.

(xiv) **Dairy Development Programmes** have also been promoted in the village Sarai Khema. There are 2 dairy development centres in the village installed individually by the village people. The dairy centres has been operational to fulfil the milk requirements of HAL and Sanjay Gandhi Hospital and Marketing Centre at Munshiganj in village Sarai Khema of the Amethi Block.

Integrated Rural Development Schemes have also been launched in the area as per recommendations of the local people's representatives. Thus rural development programmes have been given top priority in the village. Social Welfare Schemes especially scheduled castes—Special Component Plan (SCP) has been given top priority by the Government at grassroots level. It is argued that the faster development of the village is required to bridge the gap between rural to urban and urban areas. Need based rural development programmes have been implemented in the village Sarai Khema as per needs of the local village people to fulfil the increasing demand of the growing cattle and human population.

5. Conclusion

Thus it can be concluded that the development should be participatory and planned according to the local needs of the deprived rural multitudes. It should be planned through tapping local talent and resources of the area. Implementation of various rural development programmes can be made more effective through people's participation. The real meaning of democracy is the importance of the people and rural self-government.

Different activities of related line departments have been handed over to the village Panchayats and various committees have been formed to control the different rural development programmes.

MNREGA has been one of the important programme launched in the village Sarai Khema. In the MNREGA programme 498 job cards have been provided to the needy interested people of the village and ponds, chakroads etc. have been constructed as per requirement of the local people. Rural roads have also been constructed under the Pradhan Mantri Gram Sadak Yojana in the village Sarai Khema. It has been recorded that 55 low cost rural houses have been constructed for the down-trodden scheduled caste beneficiaries of the village Sarai Khema. Land reform programme has also been launched in the village. Under the rural water supply scheme 290 India Mark II hand pumps have been insalled in the village Sarai Khema of the Amethi Block. Under the rural sanitation scheme 35 rural sanitation latrines have also been constructed in the village. Most of the villagers/householders have been provided rural electrification facility including commercial electric support system for the integrated development. Various small scale and agro-industries have been established in the village Sarai Khema. Various welfare programmes for scheduled castes, backward castes, minorities, freeship and scholarships have been facilitated to the needy and interested people living under below poverty line (BPL). Rural road including pacca road, kaccha road, kharanja and chakroads have been constructed in the village Sarai Khema. Agriculture, horticulture, social forestry, animal husbandry, dairy development, animal hospital and medical health programmes are also implemented in the village Sarai Khema of Amethi Block.

Thus the rural development programmes have been implemented in the village Sarai Khema with the people participation and improper implementation of sanctioned schemes and the implementation and management of the programmes has led to the least sustainability because successful implementation of rural development programmes will lead to generate employment and increase the income of deprived and multitudes and improve the socio-economic conditions of the poor people living in the village Sarai Khema of the Amethi Block.

15

Traditional Leadership in Ghana

Jones Lewis Arthur and *Michael Kofi Nsiah*

Introduction

The links between governance, science, technology and sustainable development are multiple, complex and important. The effects on the environment, technology and development, in fact, depend on the extent to which culture, science, technology, environment, and political goals can be made complementary and mutually supportive. A positive outcome requires appropriate supporting political, economic and environmental policies at the national and international levels.

The paper examines the complexities between traditional and political leadership, science and technology and sustainable development in the context of developing countries. Particular resource information was drawn from Ghana's experience. Available literature was also reviewed. Key informants including chiefs, researchers, and community opinion leaders were interviewed on their perception on how science, technology and tradition can co-exist to demystify the issues of sustainable development in the rich cultural Ghanaian societies.

The participatory methodology was used for data collection. This included:

- Literature review or review of existing research data
- Community Forum
- Focus Group Discussions

- Interview of key informants
- Community cafes

Descriptive sampling design was used and covered a various communities of Ghana. A sample size of 500, making up of 100 households of about 5 people, were sampled for their views, comments and perceptions on indigenous knowledge, traditional leadership, environmental best practices. The sampling procedures employed included incidental, purposive, snowballing and cluster to identify floating respondents, groups and key informants through identification by friends, colleagues and neighbours. The sampled respondents included staff Chiefs and Queenmothers, traditional leadership and people living in rural communities where indigenous knowledge if primarily used. In-depth interviews were conducted with the youth and elderly people. The focus of the inquiry included views of the respondents on the advantages and dangers using indigenous knowledge and traditionalism to address environmental issues.

Data and relevant information were collected from three (3) main sources:

- Literature review—Including previous studies by researchers.
- Interview—Youth and elderly people selected from rural communities in Ghana.
- Focus group discussions—Chiefs and their elders.

It was identified that both traditional and national politics have critical roles to play in ensuring sustainable development as they stand a greater chance to demystify the misconceptions about science within the rich cultural environments of the Ghanaian society. Chiefs, Queenmothers and elders of Ghana's traditional societies are actively contributing to sustainable development through best practices, education and advocacy and drives towards a safer and productive environment and its resources. Adopting a closer bottom-to-top approach to development was identified as a pragmatic step to ensure an environmentally safe and productive development that applies science and technology to provide the needs of the present and future generations.

Introduction

Different societies have their distinctive view of nature, society, human and history. While sustainability is a global issue requiring bilateral and multilateral cooperation, the basic framework of sustainable development must embrace the distinctive social and cultural setting, with special attention to the local economic and social conditions (for example, poverty, inequality and inefficient use of resources). The issue of governance and sustainable development, therefore, demands a multiple, complex and important interface as reiterated in the UN Millennium Development Goals (MDGs).

Indigenous knowledge, leadership and biodiversity are complementary phenomena which are essential to human development. Global awareness of the crisis concerning the conservation of biodiversity was assured following the United Nations Conference on Environment and Development held in June 1992 in Rio de Janeiro. Of equal concern to many world citizens is the uncertain status of the indigenous knowledge that reflects many generations of experience and problem-solving by thousands of ethnic groups across the globe. Very little of this knowledge is being acknowledged and recorded, yet it represents an immensely valuable data base that provides humankind with insights on how numerous communities have interacted with their changing environment including its floral and faunal resources.

Wherever sustainable development decisions are taken, impacts on economic well-being, quality of life, governance and the natural environment depend on good decision support. Wrong decisions, or decisions which are blocked because of lack of public acceptability and effective collaboration with institutions including traditional societies, can have major impacts on the achievement of development goals. Sustainable development decision-making raises a wide range of environmental, social, political and economic issues apart from holding implications for the goals, objectives and development of the society concerned. Poor decisions which lack public acceptability or are not based on proper analysis can have serious impacts on the environment as well as economic and social well-being. Decision support is, therefore, absolutely critical.

Partnering traditional societies for effective governance and sustainable development is an emerging concept. Apart from the claim that traditional societies have intended purposes of promoting development, it performs to ensure the conservation and preservation of the environment for both present and future uses, key amongst them is the role of indigenous knowledge.

Generally, indigenous traditional knowledge are based on symbolism, and involve spiritual rituals, religious practices, social taboos, and sacred animal totems. Other positive practices are based on the experiential, involving travel in order to learn from the experiences of other farmers, hunters, gatherers, fishermen, herbal medicine healers, and artisans. The traditional keepers and users of local ecological knowledge and wisdom are typically the key elders from rural African communities.

Ghana has taken remarkable strides towards consolidating democratic development in recent years, with successive and successful national elections. This spans across nearly 50 years of democratic and military periods of state governance. However, the country faces a significant challenge. The rapid pace of development has led to numerous local tensions over land, resources, and chieftaincy succession, especially where traditional groups and communities have grappled with modern politics and jurisprudence. While the chieftaincy system has served as a mainstay for social cohesion in Ghana even in modern times, the politicization of some aspects of the system has contributed to these tensions. In the recent past, an absence of new mechanisms to complement traditional means for managing these tensions has led to violent conflict.

Countries the worldover, have been critical of achieving the benchmarks of sustainable development. However, institutional support for sustainable development has been inadequate as government policies have also achieved minimal success. Traditional authorities can and have valuable contributions to offer an ensuring sustainable development, however not much has been pursued. Ghana continues to lose large tracts of forest cover to environmental degradation. Domestic and industrial waste still pose problems to biodiversity and human health. Refuse continue to pile up in major cities in, including Sekondi-Takoradi, Accra, Tema and Tamale, Ghana. The issue arising has been the absence of effective policies draws on indigenous knowledge to address waste management and conservation of environmental resources.

Methodology

The participatory methodology was used for data collection. This included:

- Literature review or review of existing research data

- Community Forum
- Focus Group Discussions
- Interview of Key Informants
- Community cafes

Descriptive sampling design was used and covered various communities in the Brong Ahafo Region of Ghana. A sample size of 500, making up of 100 households of about 5 people, were sampled for their views, comments and perceptions on indigenous knowledge, traditional leadership, environmental best practices. The sampling procedures employed included incidental, purposive, snowballing and cluster to identify floating respondents, groups and key informants through identification by friends, colleagues and neighbours. The sampled respondents included staff Chiefs and Queenmothers, traditional leadership and people living in rural communities where indigenous knowledge if primarily used.

In-depth interviews were conducted with the youth and elderly people. The focus of the inquiry included views of the respondents on the advantages and dangers using indigenous knowledge and traditionalism to address environmental issues.

Complexities of Traditional Societies in Ghana

Traditional rule in Ghana finds expression in forms such as religious leadership, lineage headship, leadership in extended families, and chieftaincy. Chieftaincy is, however, the fullest expression of traditional rule in its institutionalized form. It embodies: the cardinal characteristics of prescribed kinship and lineage succession to office; awe and sacredness of office and office holders; specific forms of contractual relationships between chiefs and their subjects; and institutionalized procedures for decision-taking and implementation at the levels of local community and local participation as indicated by some selected traditional leaders in Chiraa and Abesim.

Ghana's traditional leadership structure is a hierarchy. At the base is the clan head (Abusuapannin). Next and above the clan head is the village or town chief, also known in some of the Akan dialects as the "Odikuro", literally the head of the village or the town. The next in the hierarchy is the "Omanhene" or the paramount chief, usually the traditional leader at the district level. At the apex is the head of a tribal group such as the King of Asantes or the Ga Mantse of the Gas.

The Asante dynasty, a classical example of a traditional leadership, is one of the well constituted traditional bodies in Ghana. Apart from the rather unique role of the *Asantehene,* and of the Golden Stool in welding virtually autonomous traditional states and *amanhene* together, there is very little in which the Ashanti differ from other Akans in their political set-up. Such roles are also performed by the Akyem and Ga dynasties in Ghana. More so, the Asante dynasties have clearly demarcated hierarchy of authority that transcends from the head of the clans to the Abusuapanin thus supporting the claims of Radcliffe-Brown, 1961.

Qualities and Features of Traditional Leadership

"The kingly office springs from a period in native history when there was continual warfare among the different tribes inhabiting the country. In a related incidence, the Krontihene of Sunyani, the traditional prime minister of the Paramount Chief, indicated that the choice of a king was most probably determined by the personal valour, intelligence and capability of the individual to lead the forces of the community in times of war. As he added, such an individual was undoubtedly the best man the community could produce." For example, under the kingship of his royal majesty Nana Wiafe Akenten III, the Offinso traditional council has been working to prove the quality traditional leadership offers, restoring the pride once lost and reforming into its true status. It is, therefore, apparent that traditional leaders are people imbued with good virtues as supported by many scholars.

Contributing to the debate, Casely Hayford had noted that:

> "At the head of the native state stand prominently, the chief (king) who is the chief magistrate and the chief military leader of the state. He is first in the councils of the country, and the first executive officer. His influence is only measured by the strength of his character."

Another feature of traditional rule is the sacredness of the office of the chief, and for that matter, the person of office bearers. Thus offices (and personages) are set apart from ordinary mundane phenomena including others. These in turn served to validate that the exercise of his authority was mainly through myths and tradition;

and owing to the religious-secular nature of political office, the incumbent was seen as more than human. In Ghana, this was particularly so in the cases of the *Awoamefia* of Anlo, and the *Yagbonwura* of the Gonja state. Finally, while incumbents of office could be questioned about the way the system was manipulated, rarely did people question the structure of the society and its institutions as such. In other words, one experienced enough rebellion but scarcely revolutions (Casely-Hayford J.E., 1903).

Traditional governance systems varied considerably among the different peoples that occupied the region of modern Ghana. While some groups developed very complex hierarchical structures, others had simple kin-based types. The matrilineal Akan-speaking people, for example, seemed to have evolved once of the highest forms of the complex system of governance. The Akan political system ensured socially acceptable conduct backed by general goodwill that in turn ensured social cohesion. Other groups like their Guan-speaking neighbours and the Ga-Adangme groups who originally practiced a religion-based system of government seem to have adopted and adapted the Akan model as could be inferred from the titles used for their political leaders. In the northern regions, although there are such centralized states as Mamprussi, Dagbon, Gonja and Wa that have hierarchical structures with clearly defined rules of succession and titles for their officeholders, there are many others like the Tallensi, Konkomba and Gurunsi who until fairly recently did not have such systems. It is important to note that the history of the chieftaincy institution differs among the different ethnic groups and even in the various administrative regions into which the country is divided.

The role and significance of traditional leaders have dwindled with age. People no longer continue to hold the office in an extremely high esteem as previously existed. The powers they wield have also reduced with the introduction of democratic tenants in the society. However, in many communities in the Akan societies, traditional leaders still carry a high level of respect and thus contribute substantially to the governance and development of their communities.

Traditional Leadership and Socio-Economic Development in Contemporary Ghana

Good governance and sound public management are preconditions for the implementation of sustainable development policies. These preconditions include efforts to ensure an ethical and more transparent government process, as well as decision-making practices sufficiently open to citizens. Good governance, therefore, seeks to satisfy economic, political and administrative demands of a society.

Traditional leaders play both statutory and non-Statutory Functions. They are seen contributing effectively to state governance through effective economic, political and administrative policies.

The statutory functions, as provided by respondents, among others are:

- Collection, refinement, codification and the unification of customary laws,
- Adjudication in chieftaincy disputes,
- Compilation of lines of succession to offices in the various traditional areas,
- Appointment of representations to various government statutory bodies including the Council of State, Prisons Council, National and Regional Lands Commissions and Regional Co-ordination Councils and constitutionally too, traditional rulers are barred from active partisan politics.

The non-statutory functions of traditional rulers are derived from their positions as moral/natural leaders of their respective communities. In a related incidence, a key informant interviews with selected elders in the Sunyani Municipality showed that, chiefs as influential members of their communities undertake the under-listed functions:

- Settlement of disputes through arbitration, Mobilization of their people for development purposes.
- They act as linkages between their communities and development agencies including central government departments, local government organs, NGOs, diplomatic missions, religious bodies and welfare associations and use of the agency of annual festivals when traditional rulers mobilize their people for the purpose of planning and seeking avenues and opportunities for executing development projects.

On close examination of the historical development of local

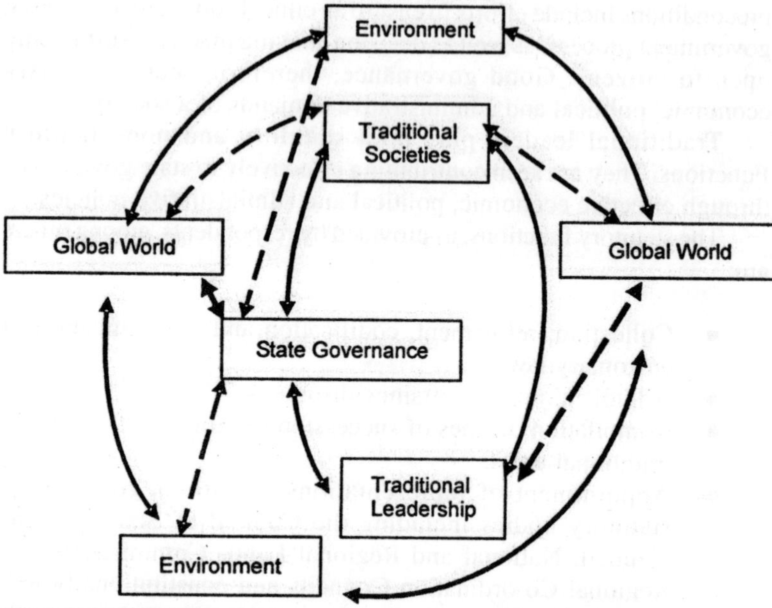

Source: Author's Construct.

Fig. 15.1: Contemporary Approaches to Sustainable Development: Exploring Critical Innovations in Traditional Leadership in Ghana

government in the country, one can discern two basic orientations. From 1951 to 1959, local government concept was conceived largely in terms of the democratic right of the local people to run their own affairs. This right consisted in the people's ability to select their own representatives who then became the policy-makers or law-givers as far as the affairs of the local areas were concerned. The right also consisted in the people's capacity to determine what services to provide, what taxes to impose, and how the proceeds of these taxes should be used. The type of local government that obtained then was akin to the British local government system with its emphasis on the principles of democracy, representation, devolution and responsibility. This was not surprising since it was from the British that Ghanaians took their cue for most of the country's political and constitutional development.

Although the traditional manners of government look awesome and fearful when seen from outside, and although the pomp and pageantry around the leaders make them look impregnable, a closer study reveals otherwise. Indeed, in many regards traditional rulers appear as loving slaves to their office. There is no movement of theirs which is not controlled by custom, tradition and taboo. In palace deliberations, for instance, the chief talks last and what he says is usually a summary of what has been said by other elders and counsellors. Besides, the traditional leader only speaks through a linguist, a titled palace official versed in customary usages and oratory. In such regards, any tendency to depart from consensus decisions will be straightened out by the linguist in the course of his presentations.

Creating the enabling functional environment for traditional leadership would result in an effective collaboration between the state and the local societies. The outflow is a society that enjoys peace, economic growth and human development. The negative impact of defective governance on economic and social development, as well as on the environment, is clear. In addition to these basic preconditions, the importance for sustainable development of key management tools such as performance measurement, mechanisms for citizen engagement, specific policy and implementation processes, and continuous strategic assessment is crucial.

In Abesim, Chiraa, Fiapre and Zongo, as environmental education programme being undertaken by the Brong Ahafo Research and Extension Centre (BAREC) in collaboration with the leadership and people of the communities proved the active contribution to Chiefs and Queenmothers to environmental education (Arthur *et al.*, 2011). Many of such Chiefs have instituted days for cleaning of their communities. They also operate functional sanitation officers that seek to the cleanliness of the people and their communities. Such communities have also contracted the services of local waste companies to clean and lift solid and liquid waste, obviously a step towards ensuring a healthy people and society.

Traditional Leadership and Local Government Nexus: The Conflicts

Democratic governance has also not succeeded in eroding such tendencies for tension between state and traditional administration. 1992 saw the swearing in into office in Ghana of an elected government. This brought into its wake the District Assemblies (DA) concept. DAs are accorded wide-ranging powers by the 1992

Constitution and the Local Government Act of 1993. Within its designated geographical area, each District Assembly is the highest political and administrative authority; planning authority; development authority; budgeting authority; and rating authority (Ayee, 2003). The District Assembly also has an overall responsibility to "co-ordinate, integrate and harmonize" the activities of all development agencies in the District inclusive of central government ministries, departments and agencies (MDAs) and non-government organisations [Article 10 (5)]. The District Assemblies work in close collaboration with the traditional leadership in areas of their jurisdiction. These chiefs serve on local boards and also on different committees within the assembly.

Ensuing an effective DA and traditional leadership, therefore, creates an interface that sometimes result in duplication of functions as well as undesirable frictions. Such complexities do mainly result from the intersecting roles and boundaries of operation, including managing waste and executing executive functions, for both state and traditional machinery of leadership. Decalo S. (1989) argues that in hoping to "modernize" their usual mono-economies, the new African leaders often espoused an "African Socialism" where the state controlled the economy. Insisting upon the need for "national integration," in the face of a plethora of ethnic collectivities, African leaders imposed a single party system, claiming that this was close to the African "palaver". However, there are also offshoots of this association that are quite undeniable. Such include the fight for control and legitimacy as well as the difficulty of respecting leadership boundaries.

In a similar view, Kwame Nkrumah had a bitter conflict with the Asantehene and other traditional leaders in Ghana who objected to being excluded from government. In Ouagadougou, a frustrated traditional emperor, the Mogho Naba of the Mossi people, attempted to use his traditional army in a quixotic attempt to dissolve an embattled Territorial Assembly. Sir Edward Mutesa II of the Baganda quarreled with Sir Andrew Cohen, Britain's last colonial governor, about the future government of Uganda and was exiled to England where he died in poverty.

In April 2000, religious and traditional leaders agreed to modify the ban in drumming and noise-making, an annual event prior to celebrating "Homowo" a festival of a large section of traditional enclave in the Greater Accra region of Ghana. The agreement was for drumming to be subdued and confined to the churches. On May

7, 2001, the first day of the 2001 ban, the Ga Traditional Council (GTC) announced that the agreement it had previously reached with local churches in 2000, which dealt on moderate and permissible noise making levels, was not applicable for 2001 and that the ban would apply to all drumming and noise-making. Christian churches claimed then that the ban was unconstitutional and that they would not observe it. Several incidents of violence were reported during the ensuing 2001 ban on drumming. In the phase of such challenges, political leadership stayed a further 'distance' away from the unfortunate developments. The resultant conflict led to massive destruction to property worth hundreds of millions of Ghanaian cedis. Clearly if there was a meaningful collaboration between these traditional societies and state governance the stalemate would not had manifested. Coherence and effective empowerment of traditional leadership, had led to many strides in effective governance, hence sustainable development. Traditional Societies are actively represented in the general assembly of the District Assemblies (DAs). Policy decisions are taken with the keen input and contributions of local authorities.

Creating an enabling interface between traditional and state leadership has not been without challenges. Leadership styles and ideologies do and have in many instances created friction in governance at both national and local levels. What confounded many western theorists on the issue were whether African leaders espoused Marxism-Leninism, African and non-African socialism, capitalism or mixed capitalism and so on, their efforts failed. A critical scenario was the situation when Dr. Kwame Nkrumah rejected compromises and ignored the advice of Sir Arthur Lewis to him, that the political-economy of the new African states should use agriculture to build their economies and should employ ethnic-based coalitions for government (Lewis, 1967). The result was that confusion reigned about how African leaders could and should deal with their economies and regimes (Lloyd *et al.*, 1970).

Partnering Traditional Leadership for Sustainable Development in Ghana

The Concept of Partnership for Sustainable Development

Partnering the traditional leadership in state governance is a vital and necessary tool for sustainable development. Key respondents including chiefs and elders of clans stated that Traditional leadership, state government and traditional societies have functional inter- and intra-relationships that impact on both local and global environmental variables. The level of partnership is as shown in (Figure 15.1).

The state, traditional leadership and societies exist in a globalized world to which there exists a close knitted collaboration. It is therefore necessary that all parties to sustainable development of human societies are integrated towards growth and development that satisfies *all* (Figure 15.1). Governance, though complex becomes easy to coordinate if considered in an integrative manner. Rooting indigenous knowledge in state governance is therefore a necessary impetus for growth and development.

Achieving sustainable development, therefore, requires an effective use of both traditional leadership and state governance since communities are mostly controlled by elders and clan heads that are answerable to chiefs and kings who also, pay allegiance to state authority.

Positioning Traditional Societies for Sustainable Development in Ghana: The Innovations

Chiefs and Kings are accorded extreme respect due to the power and authority they wield (Comments from a traditional leader in Chiraa). In the past, the authority of traditional leaders was considered way above that of colonial administrative power as well as contemporary state authority, and therefore, a critical for controlling and governing communities. Using Chiefs for development, therefore, is an active ingredient for achieving one's objective of sustainable development. Since the Chief resides with his subjects in the community, he is able to monitor and ensure that his decisions are carried out since such directives are not different from decrees. Using them, therefore, to make and monitor laws of the environment, would ensure maximum co-operation and observance of such directives.

In many traditional societies, large areas of land are reserved as sacred grooves or abodes of the gods. Observance of such directives is non-negotiable, and this ensures that the land with its resources is conserved. The belief behind the scenario is that, if such laws are not respected, one would incur the displeasure of the gods, which is mainly fatality. Respondents in Fiapre also stated that unlike Death people do not fear punishments associated with the laws of the state. Laws are sometimes flouted with impunity, since they considered non-punitive. However, the directives of Chiefs are revered and sometimes considered laws, which should be obeyed to the later. This approach, when used to regulate the use of natural resources, leads to the conservation of resources including forest ecosystems, an indigenous innovation to sustainable development.

Totems, a critical cultural practice amongst traditional societies, revolve around the use of objects as symbols of worship and reverence, are prevalent in many communities in Ghana. Animals including but not limited to the crow, deer (aboakyer festivals) are considered sacred in certain communities. It came out from respondents that, failing to protect such animals becomes a crime against the gods. Such a practice has led to the increase in the population of such animals in many communities. An interface of such idea and practice would ensure the protection of the environment and its resources.

Governance structures in Ghana are also vital as it contributes significantly to an effective way of managing the environment. Currently Ghana practices a Presidential system of government that seeks to execute power from a central government with representatives in both the regional and district administrative zones. State authority, which is mainly centrally based, makes it difficult for effective monitoring and coordination of national agenda. Making Chiefs responsive in state governance would ensure that government policies are easily accepted and observed by the various communities in Ghana. In such dimensions, an effective scenario, which seeks to promote good governance and development, would be created. Communities would also gladly accept government policies whilst working hard to ensure that they are achieved. The success story of government's massive cocoa spraying exercise for the past 4 years had been mainly due to the active and effective participation of traditional leadership and societies. Executing governance from the local levels is, therefore, the key option for championing the drive to environmental best practices.

Respondents were quick to mention that a bottom-to-up governance strategy involving local authority and central government would also encourage communities to own policies developed since they would be part of the planning and implementation of such projects. Studies have proved that community projects that infuse the suggestions and contributions of indigenes have culminated into the success of these projects. Chiefs in Ghana, though not to engage in active politics, can help better explain policies of the government whilst mobilising resources for its completion. In doing so, such Chiefs would earn the trust and support of his people which will further enhance and promote the needed governmental support for their communities. The introduction of a Kumasi Ventilated Improved Pit (KVIP) facility, a place of convenience, in most coastal communities met little support from such communities. The underlining cause was identified to be the non-involvement of such communities in the planning and execution of such projects. The people, therefore, avoided using such facilities. Through a collaborative agreement between such communities and government the people have returned to the use of these toilets since it has now been constructed at sites and in conditions accepted and approved by the communities. Sanitation and health has consequently improved in such communities. This obviously is the success story of involving state and traditional leadership in governance. Such a development, therefore, promote group living, discipline since anyone can punish as offender notwithstanding whether they are closely related. The future generation is also very critical to their present state and ability to see and survive the future. This invariably is a good impetus for growth and development of human societies.

Application of Science through Indigenous Knowledge: The Contributions of Traditional Societies

Meeting the challenges of governance in contemporary Africa is to recognize and satisfy the goals and aspirations of different groups and their leaders. Different African societies necessitated types of governance based on compromises between types of groups and individuals. International and national development agencies have recognized the value of participatory approaches to decision-making for sustainable approaches to development. During the past decade a rapidly growing set of evidence indicates a strong relationship between indigenous knowledge and sustainable development (Posey, 1985). It is, therefore, in the right perspective for state governance to actively pursue inclusive state governance that recognizes traditional leaders, societies and their contributions.

State-traditional leadership partnership is vital to resolving societal conflict as a result of the growth and development of values that are central to traditional societies. In the early 2000s, one of the longest-lasting violent conflicts in Ghana afflicted the Dagbon traditional area of the Northern region, where each of the two major clans claims affiliation to different and opposing national political parties that is, New Patriotic Party (NPP) and the National Democratic Congress (NDC). In 2003, the traditional ruler of Dagbon and a number of his followers numbering about 40 were killed in a chieftaincy dispute. Law enforcement agencies provided only a tepid response. Perceptions by the aggrieved clan of national authorities' ineptitude or compliance significantly ratcheted political tensions at the national level. This threatened to create instability and violence during national elections in December 2004. The scenario is an inducement for an effective collaboration between state and traditional leadership since its omission will mean a possible chaos. The state should recognize traditional leadership whilst providing for such their legitimate state support and discouraging the obvious involvement of traditional leadership in mainstream governance.

Traditional leadership quizzed indicated that it is not their duty to meddle in politics but rather ineffective governance of their communities and the state. Chiefs and their council of elders are, therefore, cautioned not be seen in party paraphernalia as well as partaking in party political campaigns. By respecting their boundaries in politics, it would prevent situations where they fall into the *dark side* of parties in power, such as violent campaigning which can undermine traditional authority. Meddling in state politics may result in some of the unfortunate events that led to the Dagbon crisis that occurred in Ghana. The Chief's authority is primarily vested in his people and he should, therefore, be primarily responsive to the needs of his subjects. When such a desirable interface between traditional leadership and state machinery is created, it would further enhance strides for cooperation, growth and development of human societies.

Sustainable development that seeks to provide the need of the present generation without compromising that of the future generations should be an integral component of the aspirations of societies at the community level. Chiefs, Queenmothers, though actively working towards community development, can be made more effective if given the necessary legal backing as well as state support.

Recommendations

Achieving sustainable development should see political leadership adopting an all inclusive governance system that intergrades all including and not limited to traditional leadership and societies, civil societies and development partners.

Government should strengthen traditional leadership through the regional and national houses of chiefs to ensure the promotion of peace and settlement of chieftaincy disputes. The judiciary can also be tasked to deal effectively and quickly with chieftaincy disputes in order to safeguard the chieftaincy institution. This would facilitate the existence and effectiveness of the institution. The Asantehene has instituted a policy that ensures that most chieftaincy disputes in the Ashanti kingdom are settled out of court. The current Asantehene has also requested and personally requested and ensured the withdrawal out of court of such disputes that relate to his kingdom and his sub-chiefs. This is a clear manifestation of the quest to ensure the effective participation and function of his chiefs to promote effective governance and development.

Further, development should take into account recommendations and demands of local authorities. The inputs of traditional societies should be sought before development projects are institutionalized as this would ensure a fuller cooperation of the communities. This obviously would lead to development by, for and with the people.

Effective growth and development should also seek to include the empowerment of organizations that work at the local level. Non Governmental Organizations (NGOs), Civic Groups and Community Based Organizations (CBOs) should be given adequate support to ensure their effective participation in community development. As these groups are supported, they can effectively collaborate financially with traditional leadership to execute community project that include the conservation and protection of the environment.

Governments should create the environment for unity and not align themselves with particular chieftain or people, in the societies which leads to suspicion and division. This would ensure that successive governments continue to work effectively with all communities.

Conclusion

Through the 50 years of Ghana's independence, the nation has transcended through a mirage of leadership ideologies both at local and national levels. This has passed through stages of colonial leadership that exploited traditional leadership to the present where the District Assembly concept has co-habited traditional leadership and state governance. In all such developments, the benefactors or losers have been the societies and the environment. Given the evolution of both traditional and state governance available to the societies, new strategies are needed to extend sustainable development to all societies.

State support and indigenous knowledge is recognized as a basic right that is vital to the society. Yet, governments have not been able to promote effective bottom-to-up governance that integrates the contributions of traditional leadership. With a high level of forest resource depletion, pollution of water bodies and low level of consciousness to environmental issues, it is proper and relevant that Chiefs, Queenmothers and local opinion leaders are recognized and integrated into state governance. This obviously would lead to appropriate modes of conserving and preserving the environment, hence improved well-being.

It must be noted that it is absolutely impossible to promote sustainable development of human societies without ensuring a partnership that seeks to draw traditional leadership into state governance, hence sustainable development.

The study also revealed that, Ghana has laudable policies towards dealing with environmental conservation and preservation. Such policies deal with issues including forest stock depletion, pollution and best practices. Such policies though effectively coordinated, has left much to be desired. The Ghana Standards Board (GSB), Environmental Protection Agency (EPA), Chamber of Mines have been active in the front to comprehensive and effective environmental protection and consciousness. However, their functions and policies are not effectively coordinated due to the little involvement of traditional leadership and indigenous knowledge.

REFERENCES

African Studies Quarterly, (2006). African Culture and the Problems of Government. www.web.africa.ufl.edu/asq/legal.htm (2006).

Arthur, J.L. (2006). *Health Environment Nexus—The Case of Developing Countries*, Annual Conference of Norwegian Association of Researchers (NFU 2006), University of Oslo, Norway.

Arthur, J.L., Ampadu-Sasu, K., BAREC-Ghana *et al.*, (2010). *An Overview of BAREC*, BAREC Research Conference 2010, LAP LAMBERT Academic Publishing AG and Co. KG, Germany, January 2011, ISBN: 978-3-8443-0020-8.

Ayee, J.R.A. (2003), 'Decentralisation and Local Governance in Ghana' in N. Amponsah and K. Boafo-Arthur (eds.), *Local Government in Ghana: Grassroots Participation in the 2002 Local Government Elections* (Uniflow Publishing Ltd., Accra).

Blair, Berill (2011). *Risk society on the Last Frontier: Indigenous knowledge and the politics of risk I oil resource management* at Alaska's North Slope University of Alaska, Fairbanks.

Casely-Hayford, J.E., (1903). *Gold Coast Native Institutions.* London: Sweet and Maxwell.

Cravero, K. and Kumar, C., (2005). Sustainable development through sustainable peace: conflict management in developing societies, *The Commonwealth Ministers Reference Book*, Bureau for Crisis Prevention and Recovery, United Nations Development Programme.

Decalo, S., (1989). *Psychoses of Power: African Personal Dictatorship.* Boulder, CO: Westview Press.

Ghana Web.com (2010). District Assemblies. www.ghanaweb.com/GhanaHomePage/.../district_assemblies.php (accessed 10/09/2010)

Heeds, A. (1991). *Botanical Pesticides.* Alternatives 18 (2):6-8. International Religious Freedom Report (2002): Ghana. About; Agnosticism and Atheism. http://atheism.about.com (2007)

Juan M. Pulhin, Rodel D. Lasco, Florencia B. Pulhin, Lawrence Ramos, Rose Jane J., *Community, Environment and Disaster Risk Management*, Emerald Group Publishing Limited.

Peras-Lalonde, A. (2006). *African Indigenous Knowledge and its Relevance to Sustainable Development*, IDRC Books Free online. www.idic.ca/en (2006)

Lewis, W. Arthur, (1967). *Politics in West Africa.* London: George Allen and Unwin, Ltd., pp. 42, 64.

Lloyd, P.C.; Coleman J.S. and Rosberg Jr C.G., eds., (1970). *Political Parties and National Integration in Tropical Africa.* Nigeria: Background to Nationalism. Berkeley: University of California Press, pp. 382-412.

Posey, Darrell Addison (1985a) Management of Tropical Forest Ecosystems: The Case of the Kayapo Indians of the Brazilian Amazon. *Agroforestry Systems* 3 (2): 139-158.

Public Management Committee of the OECD (2002), *Improving Policy Coherence and Integration for Sustainable Development*, OECD, Paris.

Radcliffe-Brown, A.R. (1961): "Preface". In: *African Political Systems*, eds Fortes, M. and Evans-Pritchard, E.E. London: OUP.

United Nations, 2005. *Millennium Development Goals*, New York. Goal 8 of the Millennium Development Goals (MDG) prescribes the need for nations to develop a global partnership for development. The United Nations Millennium Declaration represents a global social compact: developing countries will do more to ensure their own development, and developed countries will support them through aid, debt relief and better opportunities for trade

World Commission on Environment and Development (1987). *Our Common Future,* Oxford University Press, Oxford, p. 37.

Wilson, H.S. (1994). *African Decolonization.* London: Edward Arnold.

16

The Role of M.S.W. Students in Empowering Deprived Rural Women for Sustainable Development

H.R. Yadav

1. Introduction

The father of the nation said that the soul of India lives in villages, almost three-fourth of our population resides in rural areas. The rural people are dependent on Agriculture for sustenance. To improve inclusive and equitable growth we need to unite and integrate rural people especially the women who are the main sufferers. Most of the rural population is still deprived of Food-Cloth-Shelter (*Roti-Kapada-Makan*) common minimum requirement for their livelihood. The Government of India has stressed hard for the empowerment of rural women through generating employment/skill development, fetching the available natural and human resources, which are great challenge for inclusive sustainable development.

In view of the above Master of Social Work (MSW) students of RIWARD Post Graduate College, Hanumanganj, Sultanpur (U.P.) India, has been deployed for the fieldwork for developing "Village Information System" (VIS) for empowering deprived rural poor women through forming Self Help Groups (SHGs) and providing practical training for sustained employment for inclusive development in the various villages of Amethi Block.

Women Empowerment is a concept which is used by every-one, which indicates the increased activities for women's advancement. It has actively engaged women and others in pursuing the problem of women for decision making including programme implementation. The woman has come forward for redressing their problems on their own with the help of available support system, due to which the conditions of the women are improving but simultaneously the problems are also increasing the women's empowerment process has to overcome through a series of challenges and obstacles faced by the deprived rural poor women.

The United Nations International Conference on women held in Mexico in 1975, focused on the present activities have been started to promote the advancement and rights of the women. Such conferences have brought proposals, issues, solutions and resolutions on the different problems of the women and moved further to exert pressure on governments to listen to their problems, issues, demands, violence against women, poverty, illiteracy, women's health, reproductive rights, gender equity, women's under representation in the decision making bodies have been discussed and focused. The challenges of inequality and feminization of poverty have been unequivocally exacerbated. The women have been discussing and demanding for financing for development and specific allocation of resources for women's integrated development.

The Government of India has been stressing hard for the women empowerment through ensuring 33 per cent reservation in the three tier Panchayati Raj in situation, viz. District Panchayat, Block Panchayat and Village Panchayat to ensure political, administrative, financial management in the integrated development. Various schemes have been launched for socio-economic development of the women and the self-sustenance. The women in our country are facing wide range of biases in the society, viz. unequal, opportunities in education, employment and asset ownership. Empowering deprived rural women for sustainable development has been one of the basic task of M.S.W. students during their fieldwork and report writing.

The analysis of various indicators of development which plays an important role for policy planning and implementation of the integrated village development. Attempt has been made to analyse the natural-social-economic-institutional infrastructural indicators of the development. The progress of various ongoing implementation of rural development programmes has been described for the selected villages of Amethi Block. The village planning has also been formulated with the active participation of villagers and village representatives, with the continued involvement of village women Self Help Groups (SHGs) has been formed who have been provided practical training and demonstration for collection, storage and semi-preservation of available fruits and vegetables within the village itself. With the continued efforts and support of M.S.W. students and SHGs women beneficiaries have been practically trained for preservation and final processing of fruits and vegetables, packaging and marketing of finished products. Preservation and processing of fruits and vegetables have been an asset for controlling the perishability of fruits and vegetables because it provided appropriate price to the horticulture growers and generated continued employment to the deprived rural women Self Help Groups (SHGs) beneficiaries which increased their income and also improved their socio-economic conditions.

2. Objectives of the Study

The main objectives of the study are as follows:

- To describe the indicators of planning and development of the village.
- To analyse the progress of rural development programmes implementation.
- To develop village information system.
- To explain the village planning formulation with the participation of villagers and MSW students.
- To explain the formation of Self Help Groups credit and banking support for SHGs.
- To describe the preservation processing of fruits and vegetables packaging, marketing, training and demonstration to the women Self Help Groups in the various villages of Amethi.
- To analyze the Techno-economic viability of the training and demonstration of the self-sustenance of the Self Help Groups women beneficiaries for inclusive development.

3. Indicators of Village Development Planning

The Master of Social Work (MSW) Students selected various natural-social-economic-institutional-infrastructureal major indicators for developing village planning (see Fig. 16.1 and Fig. 16.2).

Natural	Social	Economics	Institutional	Infrastructural
↓	↓	↓	↓	↓
Physiography Drainage System Rainfall Soil Land use Forest and vegetation surface water underground water.	Population caste Workers Education Religion Welfare Social Security Employment	Agriculture Horticulture Social- Forestry Land ownership Irrigation Productivity Animal Husbandry Primary- Secondary and Tartiary Sector of Economy	Education Administration Health Communication Rural Development Agriculture Reverence Financial/Banking Industry Irrigation.	Religious and Historical Administrative Irrigation Road communication drinking water energy medical health Agro-Industry Smell scale industry Hospital Banking Education

Fig. 16.1: Indicators of Village Development Planning

I. **Natural Resources:** The villagers Khasara Map has been taken as a base map to analyze the natural resources of the village including cropping quality land physiography of the village, drainage system, rainfall, soil, land use, forest and vegetation, surface and underground water resources available in the villages of Amethi Block. Land-Sat-Imagery Data has also been made available to the students. Geo-referencing of village Khasara Map, Survey of India Toposheets, and Land-Sat-Imagery maps available upto 1:10,000 scale has been done adopting geographic information system and the available data has been digitized for appropriate analysis of available natural resources.

II. **Social Resources:** The M.S.W. students are not only relying on the secondary sources of information but their study is based on the primary sources of data collection with the involvement of peoples representatives, village administrative official for the collection and collation of population (Male-Female-Children), caste comparison, workers (Primary-

Fig. 16.2: Village Map of Amethi Block

Secondary-Tertiary), Educational profiles etc. Religion, welfare programme, social security and employment stature of village level.

III. Economic Resources: The informations regarding economic resources are being collected for cultivation, agricultural, production, Animal husbandry, primary-secondary-tertiary sectors of economy at village in Amethi Block has been collected, collated to ensure village development planning.

IV. Institutional Resources: The institutional resources information have been collected for historical religious institutions, Education, administrative, health, communication, rural development agriculture, revenue, financial-banking-cooperatives industry and irrigation institutional resources at village level data has been analysed for the formulation of village development planning in Amethi Block.

V. **Infrastructural Resources:** The infrastructural resources data has been collected at village level which includes—historical, religious, infrastructure, administrative, irrigation, road, communication, education, drinking water, energy, medical/health, agro-industry, small scale industry, animal hospital, banking, social welfare programmes, because the available infrastructure resources information will play an important role information of village development planning.

4. Progress of Rural Development Programme

Various rural development programmes are being implemented at village level which includes mainly—NAREGA, Energy Small Scale Industry, Dinking Water, Sanitation, National Rural Health Mission (NRHM), Women and Child Development/Women Empowerment schemes, Agriculture, Horticulture, Social Forestry, Animal Husbandry/ Dairy Development, Medical Health Education, Employment Generation etc. are being implemented by various implementing agencies which are getting financial support from various sources including Central Government, State Government and different funding schemes. The progress of the implementation of various rural development schemes and its financial flow from various sources are depicted in chart (Fig. 16.3).

5. Developing Village Information System

Decentralized micro (village) level development planning and implementation of need based various rural development programmes with the people's participation is only possible when there is an appropriate data base for natural and human resources are available at village level and the progress of various ongoing rural development programmes etc. are available at village level, then only need based action plan with people's participation can be formulated for its implementation and monitoring of various rural development programmes at village-block district level, paper collection of institutional, infrastructural, socio-economic data base at village level and mapping of natural resources should be developed at micro-level

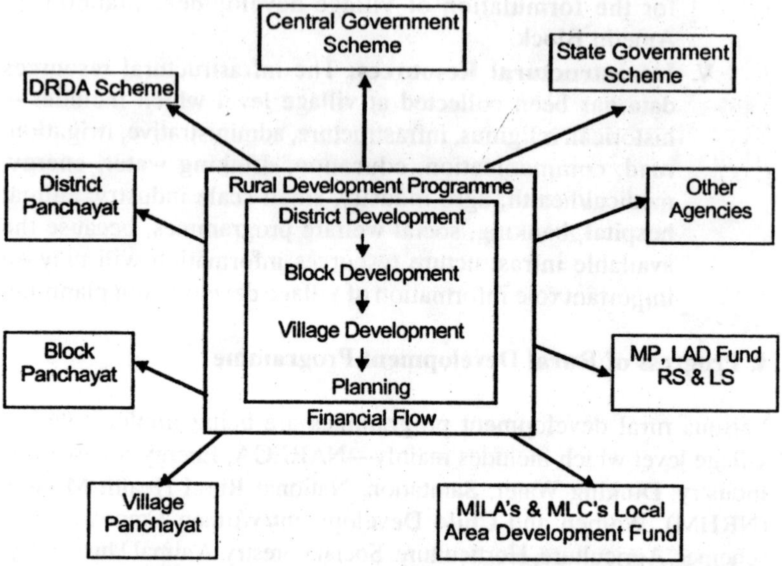

Fig. 16.3

and establishment of Village Information System (VIS) for the village people to ensure need based village action plan will be an asset to strengthen **village → block → district Level Panchayati Raj Institutions.** The **Village Information System** will be an asset to strengthen the three-tier Panchayati Raj system and to participate in the implementation of planned developmental activities to avail the fruits, services and benefits made available by Central and State Governments and other agencies. The important line departments are mentioned in the Chart.

Strengthening of Panchayati Raj Institutions is possible when thee is an appropriate Village Information System, and need of the people is identified, accordingly village plan is formulation with the continued involvement of people and the planned programmes are being implemented through the people's participation and proper follow up and monitoring is also ensured by the local involved people.

There are various implementing agencies and institutions engaged in the implementation of various rural development programmes at village level which are depicted in the chart (Figs. 16.4 to 16.6).

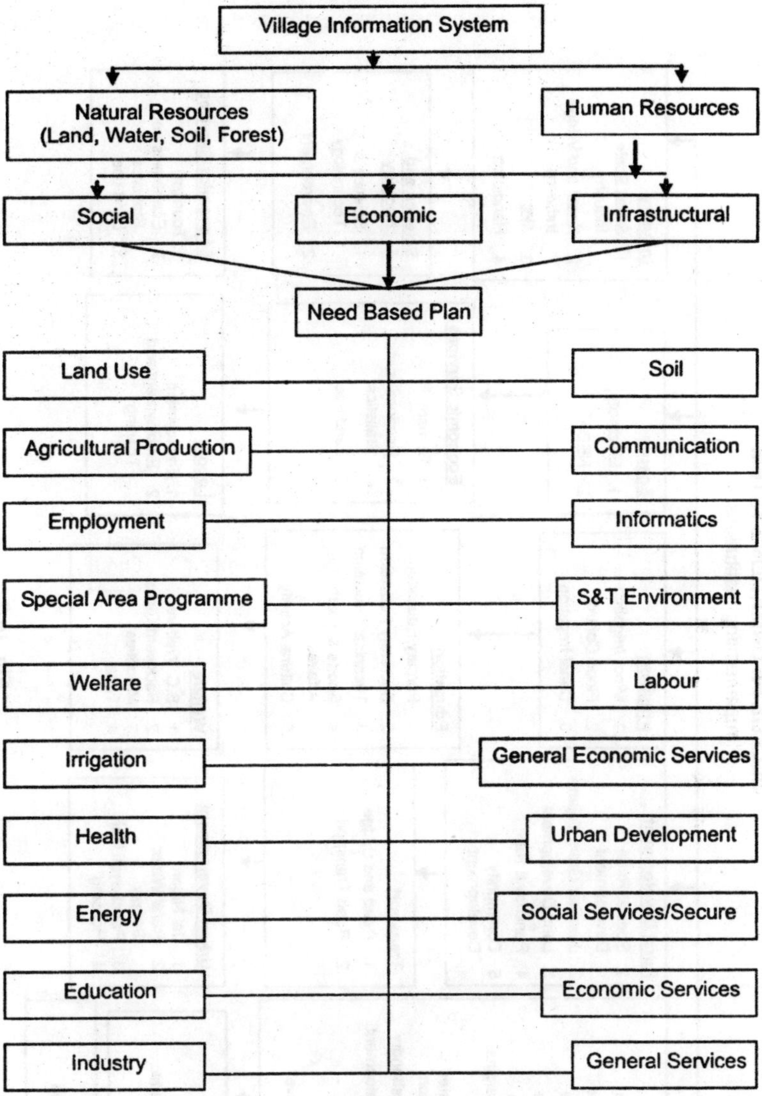

Fig. 16.4

Various sources of data for village information system for village development planning is depicted in chart which shows the availability of data from secondary sources and also the collection of data at ground level, i.e. primary survey. The Geo-referencing of data base

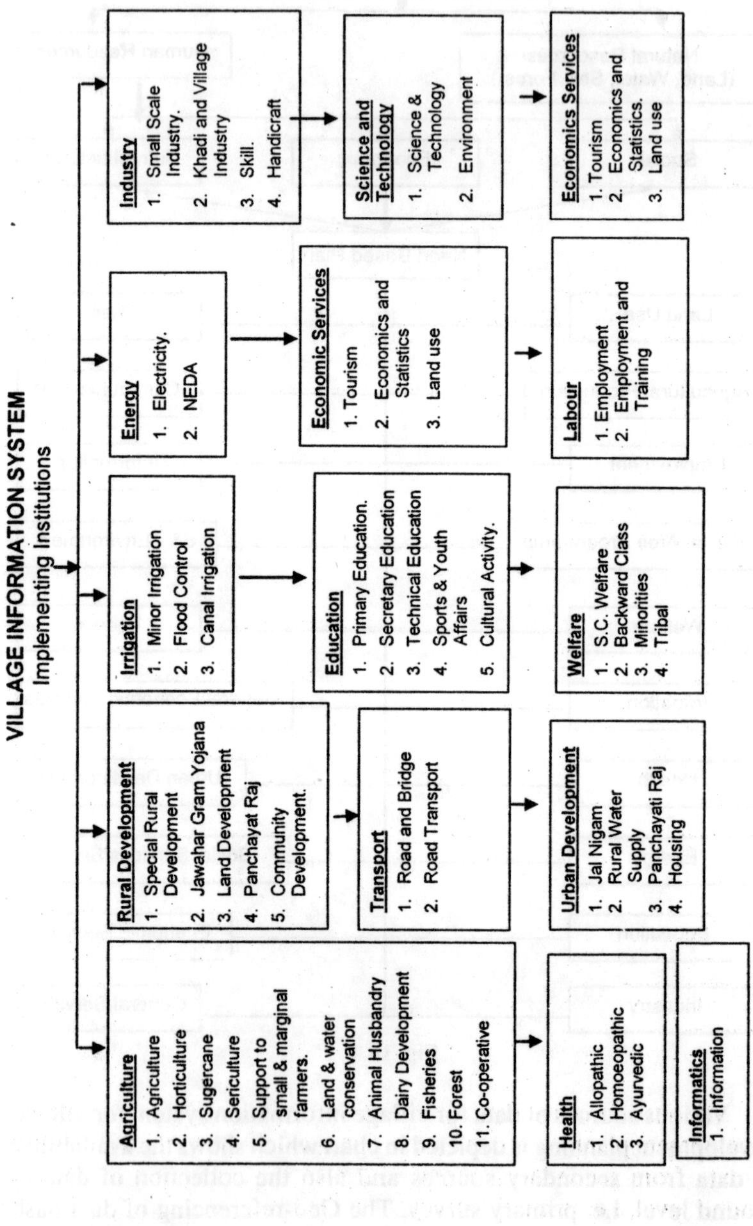

VILLAGE INFORMATION SYSTEM
Implementing Institutions
Agriculture
1. Agriculture
2. Horticulture
3. Sugercane
4. Sericulture
5. Support to small & marginal farmers.
6. Land & water conservation
7. Animal Husbandry
8. Dairy Development
9. Fisheries
10. Forest
11. Co-operative
Rural Development
1. Special Rural Development
2. Jawahar Gram Yojana
3. Land Development
4. Panchayat Raj
5. Community Development.
Irrigation
1. Minor Irrigation
2. Flood Control
3. Canal Irrigation
Energy
1. Electricity.
2. NEDA
Industry
1. Small Scale Industry.
2. Khadi and Village Industry
3. Skill.
4. Handicraft
Health
1. Allopathic
2. Homoeopathic
3. Ayurvedic
Informatics
1. Information
Transport
1. Road and Bridge
2. Road Transport
Education
1. Primary Education.
2. Secretary Education
3. Technical Education
4. Sports & Youth Affairs
5. Cultural Activity.
Economic Services
1. Tourism
2. Economics and Statistics
3. Land use
Science and Technology
1. Science & Technology
2. Environment
Urban Development
1. Jal Nigam
2. Rural Water Supply
3. Panchayati Raj
4. Housing
Welfare
1. S.C. Welfare
2. Backward Class
3. Minorities
4. Tribal
Labour
1. Employment
2. Employment and Training
Economics Services
1. Tourism
2. Economics and Statistics.
3. Land use

Fig. 16.5

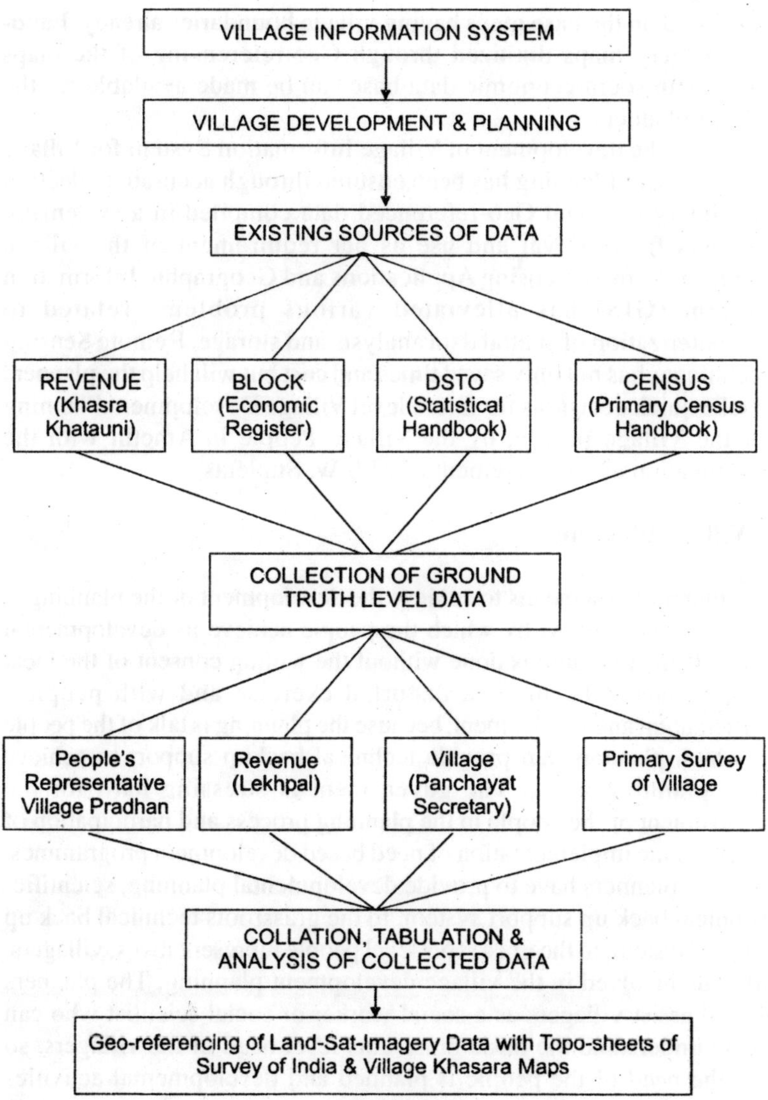

Fig. 16.6

and digitization of Land-Sat-Imagery data through Geographic Information System (GIS) is being analysed with the support of Department of Science and Technology, Government of India, New Delhi. The collected data and informations in ground is further integrated on the base maps having village boundaries already, Land-Sat-Imagery maps digitized through Geo-referencing of the maps along with socio-economic data base can be made available for the village planners.

Thus the development of Village Information System for Village Development Planning has been ensured through accurate collection of primary data and Geo-referenced data compiled in a systematic manners for retrieval and use as per requirement of the village planners. Remote Sensing Applications and Geographic Information System (GIS) has alleviated various problems related to computerization of spatial data analysis and storage. Remote Sensing Techniques has not only saved times and cost but will help the planners to take quick decisions for micro level Village Development Planning for the Village people, by the Village People in Amethi with the modification and involvement of M.S.W. Students.

6. Village Planning

The planning is a means to achieve the development or the planning is an organized process by which the people achieve its developmental goals. If the planning is done without the willing consent of the local people then it becomes a distorted exercise and with people's participation and involvement, because the planning is talk of the people in which planners can provide technical backup support to achieve their planning goals. The government is stressing hard for the involvement of the people in the planning process and participation of people in the implementation of need based development programmes.

The planners have to provide developmental planning, scientific-technical back up support system, to the grassroots technical back up support system, to the grassroots level people representatives, villagers, officials involved in the village development planning. The planners should assist villagers as a social worker or social scientist who can easily understand the basic grassroots level need of the villagers, so that the need of the people is planned and developmental activities are implemented with the participating of rural deprived village people to ensure to achieve the planned target of the village development.

The development should be participatory, tapping the local talent and resources in order to be sustainable village is the development unit in the new system of decentralization. Local development will be done by the local people, real administration is transferred to the local rural village unit in which planning development works, its management and control is ensured through peoples participation.

Planning will be done from bottom to top. Revolutionary step of decentralization will have complete change among the villagers life and it will have new era of integrated village development.

In view of the decentralization of administrative and financial powers of Village-Block-District Panchayat; the powers have been delegated to the three-tier Panchayati Raj Institutions. Village Panchayats have been handed over powers on land management, Education Government Tube wells, Hand Pump, Youth Welfare, Medical Health, Women and Child Development, Animal, Food supply shops, Agriculture rural development, Panchayati Raj, Pension and scholarship programme at village level. There are five standing committees of the Panchayats, viz. Planning and development, Education, Construction work, Health and Welfare and Administrative Committees are responsible for the formulation of plan and implementation of planned developmental activities maintenance and sustenance of the completed projects.

In view of the **"Self-Government"** of Panchayat at village level in Amethi Block need based people's participation action plan has been formulated to fulfil the dreams of "Self-Government", i.e. **Panchayati Raj** to **Lok Raj** to **Gram Raj** and to establish the **Ram Raj** which had been the dream and **vision of the Mahatma Gandhi.**

Need based village level planning is being done through the MSW students as a Social Worker in the villages of Amethi in which Land Management and Land Development, wastelands development, agriculture, horticulture, social forestry, agro-industries, gramodyog, employment, India Mark II Hand pumps, road, communication, energy (Electrification), animal, welfare, education, medical and health women and child development, youth welfare, Government tube-wells, food supply shops, rural development, Panchayati Raj, pension and scholarships etc. have been planned with the involvement of peoples representative and villages.

7. Formation of Self Help Groups (SHGs)

With the active participation of Master of Social Work students, village information system has been developed and village development planning has also been formulated with the active participation of grassroots level village beneficiaries, the students have been advised to select deprived, resourceless below poverty line eleven beneficiaries where Self Help Groups (SHGs) have been formed.

The rural poor women are the main sufferers in the villages who are deprived of basic amenities required for proper livelihood. To empower rural women Self Help Groups (SHGs) have been formed. Self Help Groups are small informal association created for the purpose of enabling the members to reap the economic benefits out of mutual help, solidarity and joint responsibility.

These homogeneous small groups are involved in saving and credit activities are capable of taking care of disks through monitoring. Therefore, strengthening the common Self Help Groups (SHGs) beneficiaries for self-sustenance through collection and storage of excess available fruits and vegetables, preservation and processing of fruits and vegetables packaging and marketing for value addition, employment generation and increasing the income of deprived rural poor women to ensure women empowerment for inclusive development.

MSW students has successfully formed 30 Self Help Groups (SHGs) in the different villages of Amethi. The SHGs have also been open Bank Account in the Baroda U.P. Gramin Bank at Tala, Amethi.

Grants and C.C.L. amount including the amount as on 31.3.2009 in their Bank Account is mentioned in Table 16.1 which indicates the willingness and interest of women SHGs for their self-sustenance for inclusive development.

8. Preservation and Processing of Fruits and Vegetables Packaging and Marketing

Initially the MSW students have been given practical training and demonstration for preservation of fruits and vegetables, it packaging and marketing in the F.P.O. based laboratory at Sultanpur. The MSW students along with the scientific trainer-cum-instructor practically demonstrated and trained the storage, collection, cleaning of available fruits and vegetables, with the view to control and check the perishability of excess available fruits and vegetables. Final preservation and processing of fruits and vegetables have been demonstrated to the women Self Help Groups (SHGs) beneficiaries packaging and marketing of the finished product have also been practically demonstrated to the selected women Self Help Groups (SHGs) beneficiaries in the various villages of Amethi Block.

Table 16.1: Women Beneficiaries Self Help Group (SHGs) Bank Account No. and Amount and C.C.L. Amount

Sl. No.	Names of S.H.Gs.	Address	Chairman of S.H.Gs	Mem-bers of SHGs	Baroda U.P. Gramin Bank, Tala, Amethi, Sultanpur		
					A/c No.	C.C.L (Rs.)	Amount as on 31.3.09
(1)	(2)	(3)	(4)	(5)	(6)	(7)	(8)
1.	Saraswati SHG	Semrauta, Amethi, Sultanpur.	Smt. Israwati	12	8164	70,000	1370
2.	Ujala SHG	Kushi Tali Amethi, Sultanpur.	Smt. Nisha	11	8176	—	1024
3.	Arati SHG	Kushi Tali Amethi, Sultanpur.	Smt. Sumitra	11	8716	—	1560
4.	Jaimata Di SHG	Kushi Tali Amethi, Sultanpur.	Smt. Mamta	13	8715	80,000	4635
5.	Laxmi SHG	Kushi Tali Amethi, Sultanpur.	Smt. Shanti	12	8165	—	1985
6.	Jyto SHG	Loharta, Amethi, Sultanpur.	Smt. Pratima	12	6878	37000	251
7.	Bishal SHG	Maha Dewan Tala, Amethi, Sultanpur.	Smt. Anar Kali	10	7276	31000	1314
8.	Nirmal SHG	Aasare Tewari Ka Poora, Amethi, Sultanpur.	Smt. Nisha Devi	12	9396	—	4968
9.	Jai Ma Durge SHG	Somepur, Mankanth Amethi, Sultanpur.	Smt. Manju Devi	12	8080	86000	2981
10.	Vikas SHG	Poore Aasare Tiwari Ka Poora, Amethi, Sultanpur.	Smt. Bimla Devi	12	7415	37000	1959
11.	Ambedkar SHG	Poore Shivlal Amethi, Sultanpur.	Smt. Sangeeta	10	8146	43000	1360
12.	Neelam SHG	Poore Aasre Tiwari Ka Poora, Amethi, Sultanpur.	Meena Devi	10	7425	28000	3030

(Contd.)

Table 16.1 (*Contd.*)

(1)	(2)	(3)	(4)	(5)	(6)	(7)	(8)
13.	Saurabh SHG	Jangal Tikari, Amethi, Sultanpur.	Smt. Sarita	11	7785	79000	5099
14.	Shanti SHG	Loharata, Amethi, Sultanpur.	Smt. Sudha Rani	8	8090	-	3160
15.	Jagriti SHG	Poore Galkata Tala, Amethi, Sultanpur.	Smt. Nirmala Devi	11	8114	86000	998
16.	Vikas SHG	Poore Durga Bhaganpur, Amethi, Sultanpur.	Smt. Pushpa	12	8148	-	200
17.	Eeshwar SHG	Poore Ichcha, Amethi, Sultanpur.	Smt. Reeta	10	8042 Alld.B	-	2610
18.	Laxmi SHG	Poore Ichcha, Amethi, Sultanpur.	Smt. Sarita	13	8017 Alld.B	-	2325
19.	Yadubansi SHG	Bhaganpur, Amethi, Sultanpur.	Smt. Raj Kali	11	8742	-	1305
20.	Laxmi SHG	Umapur Ganapatti, Amethi, Sultanpur.	Smt. Meena	13	8142 Alld.B	-	2935
21.	Jai Ma Durga SHG	Bhaganpur, Amethi, Sultanpur.	Smt. Mithilesh	10	8157	-	1040
22.	Manshal SHG	Poore Mogh, Amethi, Sultanpur.	Smt. Genda	13	8725	-	1320
23.	Jyoti SHG	Darka Deeh, Darka, Amethi, Sultanpur.	Smt. Sapna	12	7748	43000	4590
24.	Hari SHG	Poore Aasre Tiwari, Amethi, Sultanpur.	Smt. Geeta	11	8079	-	1190
25.	Duga SHG	Dhamrawana, Jangal Tikari, Amethi, Sultanpur.	Smt. Champa	12	8376	-	1210
26.	Gandhi SHG	Poore Shiv Gulam, Amethi, Sultanpur.	Smt. Kiran Devi	12	8084	28000	3970

(*Contd.*)

Table 16.1 (*Contd.*)

(1)	(2)	(3)	(4)	(5)	(6)	(7)	(8)
27.	Shiv SHG	Poore Ram Lal, Amethi, Sultanpur.	Smt. Ranjana	12	7173	-	2640
28.	Vaishno SHG	Jangal Tikari, Amethi, Sultanpur.	Smt. Geeta	12	7784	79000	3731
29.	Shiv Shankar SHG	Jangal Tikari, Amethi, Sultanpur.	Smt. Phool Kala	13	8994	-	1550
30.	Om Nama Shiway SHG	Dhamrawana, Jangal Tikari, Amethi, Sultanpur.	Smt. Sarita Devi	13	9306	-	1950

The preservation and processing of fruits and vegetables have been an asset to the horticulture growers to check the damage/loss and perishability of fruits and vegetables and it has been also and asset in providing appropriate price to the horticulture producers.

The collection cleaning and storage of available fruits and vegetables generated an employment to the deprived rural poor women Self Help Groups (SHGs) beneficiaries and increased their income and improved their socio-economic conditions for sustainable inclusive development.

The preservation and processing of fruits and vegetables training and demonstration to the selected women Self Help Groups (SHGs) beneficiaries has been described in seven phases of programme implementation with the involvement of women Self Help Groups (SHGs).

1st Phase: It will be a value addition to the fruits and vegetables growers, semi-preservation of fruits and vegetables will control and check the perishability of fruits and vegetables, hence ensuring value addition and appropriate price for their horticultural produce.

2nd Phase: The collection, cleaning, storage and semi-preservation of fruits and vegetables will make the horticulture growers of sustaining through employment and income generation.

3rd Phase: Semi-preservation and Semi-processing of fruits and vegetables will generate employment and increase the income of the involved women Self Help Groups (SHGs) beneficiaries hence it will ensure the self-sustainability.

4th Phase: The final fruits and vegetables will also generate employment and increase the income of deprived and poor women Self Help Groups (SHGs) beneficiaries due to which it will become self-sustainable.

5th Phase: The packaging and marketing of the finished products will generate employment and increase their income, subsequently it will ensure self-sustainability.

6th Phase: Since the Training-cum-Production Centre is being maintained and managed with the, by the, for the women Self Help Groups (SHGs) beneficiaries automatically the project will become self-sustainable.

7th Phase: In an integrated manner starting from Horticulture growers to involved women Self Help Groups (SHGs) beneficiaries up to the marketing and sharing of accrued benefits. Each and every involved involved beneficiary will be benefited, therefore, the project implementation will become self-sustainable and viable.

9. Techno-Economic Viability

The collection, cleaning, storage, preservation and processing of fruits and vegetables will provide value addition and support price to the horticulture producers and the preservation and processing of fruits and vegetables, packaging and marketing of the finished products will generate employment and increase the income of the involved women Self Help Groups (SHGs) beneficiaries which has become self-sustainable because it has improved the socio-economic conditions of the involved women Self Help Groups (SHGs) beneficiaries cooperative has also been formed to ensure proper packaging and marketing of the finished products.

The development of Village Information System (VIS) and formulation of village development planning and formation of Self Help Groups of the selected women beneficiaries and collection, cleaning, storage, preservation and processing of fruits and vegetables and its packaging and marketing and sharing of accrued benefits will be technically and economically viable. The techno-economic viability of the programme has been depicted in the chart (Fig. 16.7).

10. Conclusion

It is known fact that the youths are i-kon and especially the students are future, students manpower especially during their study and field-work, the student can achieve the national goal. Realizing the power of the M.S.W. students, they have been allotted one village of the Amethi Block, in which they collected the information related to the Register village khasara map and tried to analyse the natural land use, soil, water and cropping quality of land of the selected village.

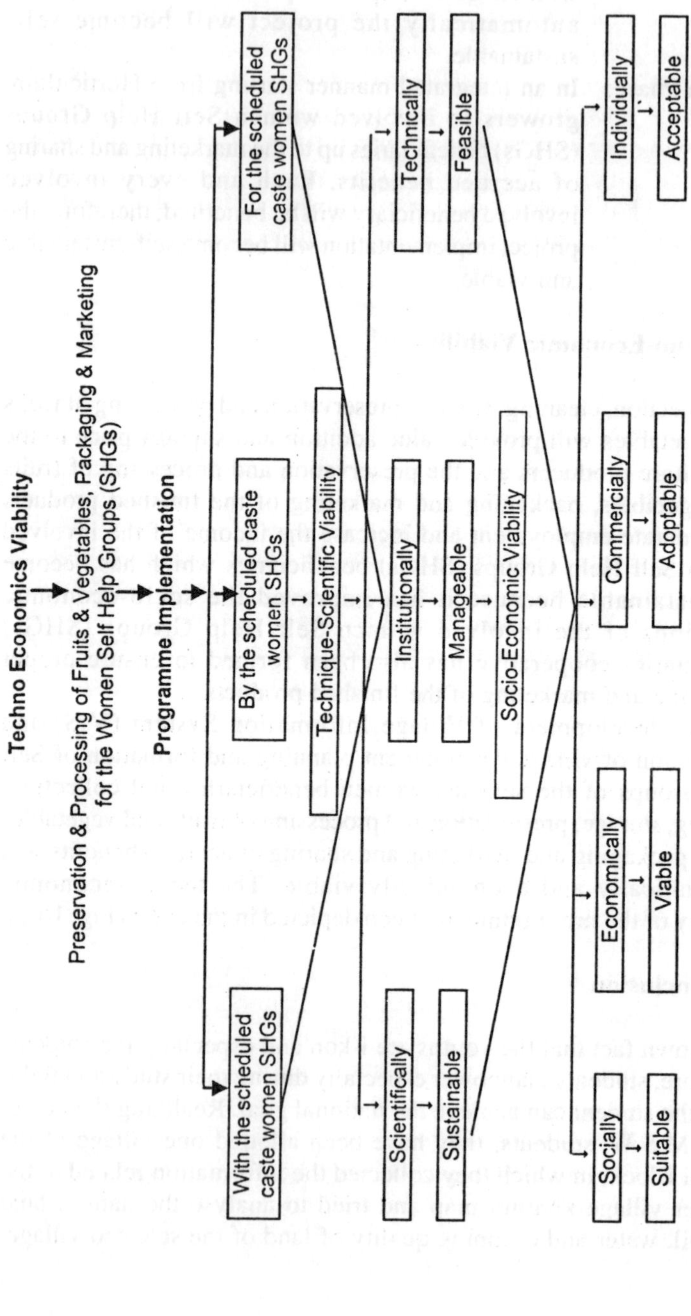

Techno Economics Viability
Preservation & Processing of Fruits & Vegetables, Packaging & Marketing for the Women Self Help Groups (SHGs)
Programme Implementation
With the scheduled caste women SHGs
By the scheduled caste women SHGs
For the scheduled caste women SHGs
Technique-Scientific Viability
Scientifically
Sustainable
Institutionally
Manageable
Technically
Feasible
Socio-Economic Viability
Socially
Suitable
Economically
Viable
Commercially
Adoptable
Individually
Acceptable

Fig. 16.7

Students also collected information through Primary survey for the social, economic, institutional infrastructural data, which has been collated and Geo-referencing of the natural and social-economic data has been carried.

Thus the students tried to collect natural, social, economic, institutional and infrastructural informations at village level in Amethi. M.S.W. students also collected information regarding progress status of the on going implementation of various rural development programmes.

The Village Information System (VIS) has been developed which has been helpful in the formation of need base village development planning with the continued involvement and participation of village peoples representatives and interested villagers.

The M.S.W. students also selected deprived, resourceless, poor 11 women beneficiaries in the selected villages of Amethi. Self Help Groups (SHGs) of selected women beneficiaries have been formed and their Bank Accounts have been opened with the view to provide banking support for their survival. The women Self Help Groups (SHGs) beneficiaries have also been provided grants-in-aid and Cost Credit Limits (CCL) facility as per norms of the bank.

Initially M.S.W. students have been trained practically for the preservation and processing of fruits and vegetables, packaging and marketing of the finished products. The experts and the students imparted practical training and demonstration to the selected women Self Help Groups (SHGs) beneficiaries for the collection, cleaning, storage, preservation and processing of available fruits and vegetables in their respective villages, its packaging and marketing has also been ensured with the help of the concerned students and the accrued benefits have been properly distributed among the women Self Help Groups (SHGs) beneficiaries as per their participation and involvement.

Thus the development of Village Information System (VIS) has been an asset in the formulation of the need based Village Development Planning to meet the villagers common minimum needs and also to strengthen the three-tier Panchayati Raj System from **Gram Raj** to **Lok Raj.**

The preservation and processing of excess available fruits and vegetables collection, cleaning and storage and preservation and processing of fruits and vegetables, packaging and marketing has been helpful in controlling/checking the damage/loss and perishability of excess produced fruits and vegetables and it provided value addition and appropriate price to the horticulture and growers for the produced fruits and vegetables.

The involved women Self Help Groups (SHGs) beneficiaries also obtained employment through the preservation and processing of fruits and vegetables, packaging and marketing of finished products, which have not only generated the employment but also increased the income of deprived rural poor women Self Help Groups (SHGs) beneficiaries and it also improved the socio-economic conditions of the villagers. Therefore, it can be concluded that the involvement and participation of M.S.W. students in the development of Village Information System to formulate need based Village Development Planning and formation of the women Self Help Groups (SHGs) beneficiaries and imparting practical training and demonstration for preservation and processing of fruits and vegetables, it is packaging and marketing, has generated employment of increased the income of deprived rural poor women beneficiaries and it has been an asset in improving their socio-economic conditions for integrated sustainable inclusive development of the villagers of Amethi.

17

Development of India through Integrated Rural Development

Vivek Kumar Singh

In the Indian context rural development assumes greater significance as 72.22 per cent (according to the 2001 census) of its population still live in rural areas. Most of the people living in rural areas draw their livelihood from agricultural and allied sectors (60.41% of total workforce), and poverty mostly persists here (27.1% in 1999-2000). At the time of independence around 83 per cent of the Indian population was living in rural areas. Accordingly, from the very beginning, our planned strategy emphasized rural development and will continue to do so in future. Strategically, the focus of our planning was to improve the economic and social conditions of the underprivileged sections of rural society. Thus, economic growth with social justice became the proclaimed objective of the planning process under rural development. It began with an emphasis on agricultural production and consequently, expanded to promote productive employment opportunities for rural masses, especially the poor, by integrating production, infrastructure, human resource and institutional development measures.

There are no universally accepted approaches to rural development. It is a choice influenced by the time, space and culture. The term development connotes overall development of rural areas to improve the quality of life of rural people. In this sense, it is a comprehensive and multidimensional concept, and encompasses the development of agriculture and allied activities, village and cottage industries and crafts, socio-economic infrastructure, community services and facilities and, above all, human resources in rural areas.

As a phenomenon, rural development is the end-result of interactions between various physical, technological, economic, social, cultural and institutional factors. As a strategy, it is designed to improve the economic and social well-being of a specific group of people—the rural poor. As a discipline, it is multi-disciplinary in nature, representing an intersection of agriculture, social, behavioural, engineering and management sciences.

The Back-Drop

Public policies and strategies of development in India are well reflected in the very preamble of the Constitution which the country adopted after independence. It provides an insight into the kind of society the founding fathers had desired and were determined to build. As the basic law of the country, the Constitution has taken steps in the direction of eliminating some of the prevalent injustices and inequities of the socio-economic system. It not only declared the right of every citizen "to work, to education and to public assistance in case of unemployment, old age, sickness and disablement, and in other cases of undeserved want", but also made a number of provisions for the advancement of socially and educationally backward classes of citizens and for the Scheduled Castes and Scheduled Tribes. It also made special provisions for women and children.

The Planning Commission was set up in March 1950 soon after the Constitution became operative. The Government resolution for establishing the Commission described the objective of planning as promotion of a rapid rise in the standard of living of the people by efficient exploitation of the resources of the country, increasing production and offering opportunities to all for employment in the service of the community.

A short introduction to the First Five Year Plan presented in December 1951 throws further light on the policies and strategies of development in India. Under the sub-head, 'Democratic Planning', it posed: "What kind of an economic system we need to achieve our aims?", and answered: "The prevailing inequality, economic stagnation and poverty indicates the necessity for change." After surveying the various systems, it was clear that "of all the alternatives available to ensure justice—social, economic and political, the establishment of a democratic polity, in the context of Indian situation, was the best." It stated further "Economic progress under democratic planning is perhaps less spectacular, but surely more enduring. It is achieved through the willing sacrifice and cooperation of the people. To carry out a plan in a democratic state implies hard work and participation by every individual. It is a process of education which, while it secures the progress of the country, also raises the quality of the individual" (Government of India, 1951:14).

This has, thus, provided the framework for formulating broad policies, programmes and strategy of planned development of India. In fact, the process of development planning had started much earlier than the commencement of First Five Year Plan in 1951. Development planning had been envisaged by the India National Congress right from the thirties. Probably, no other developing country had, for so long a period, looked forward with so sure a conviction that it would shoulder the responsibility for planning the course of its own development. This pre-independence planning took a definite shape with the appointment of the National Planning Committee in 1937 when national governments were installed in a number of provinces (States). Several sub-committees including one for rural development were set up by the Committee; and vigorous activities of all-round development were initiated even though limited in scope and coverage.

The First Five Year Plan initiated a process of all-round, balanced development to ensure a rising national income and a steady improvement in living standards over a period. It was also emphasized that the aim was not merely to plan within the existing socio-economic framework but to change the framework progressively and by democratic methods in keeping with the larger ends of the basic principles enunciated in the Constitution. It was, thus, a step in new direction towards greater assumption of direct responsibility by the State for development and a higher level of coordination in all economic sectors and at levels.

In 1954, the Parliament adopted a resolution stating that the objective of development should be the establishment of a Socialistic Pattern of Society. The term 'Socialistic Pattern of Society' was elaborated in the Second Five Year document with its long-term strategy for economic and social development. It states, "Essentially, this means that the basic criterion for determining lines of advance must not be private profit but social gain, and that the pattern of development and the structure of socio-economic relations should be so planned that they result not only in appreciable increase in national income and employment but also in greater equality in income and employment but also in greater equality in incomes and wealth." This approach "is not roared in any doctrine of dogma. Each country has to develop according to its own genius and traditions. The directive principles of state policy in the Constitution have indicated the approach in broad terms.... Economic policy and institutional changes have to be planned in a manner that would secure economic advance along democratic and egalitarian lines."

Developement Programmes

- Integrated Rural Development Programme (IRDP)
- Training of Rural Youth for Self Employment (TRYSEM)
- Supply of Improved Tool-kits to Rural Artisans (SITRA)
- Development of Women and Children in Rural Areas (DWCRA)
- Ganga Kalyan Yojana (GKY)
- Million Wells Scheme (MWS)
- Indira Awas Yojana (IAY)
- Jawahar Rojagar Yojana (JRY)
- Intensified JRY (IJRY)
- Operation Black Board (OBB)
- Employement Assurance Scheme (EAS)
- Krushak Kalyan Karyakarma (KKK)
- Rural Connectivity Programme (RCP)
- Swarnajayanti Grama Swarojagar Yojana (SGSY)
- Jawahar Gram Samridhi Yojana (JGSY)
- Drought Prone Area Programme (DPAP)

Integrated Rural Development Programme (IRDP): IRDP launched on October 2nd, 1980 all over the country and accordingly all the 15 Blocks of Boudh-Kandhamal district have been covered under the Scheme. Since then, prior to the above period, IRDP was in operation in 8 blocks of the district since 1978-79. The IRDP continues to be a major poverty alleviation programme in the field of Rural Development. The objective of IRDP is to enable identified rural poor families to cross the poverty line by providing productive assets and inputs to the target groups. The assets which could be in primary, secondary or tertiary sector are provided through financial assistance in the form of subsidy by the Government and and term credit advanced by financial institutions. The programme is implemented in all the blocks in the country as a centrally sponsored scheme funded on 50:50 basis by the Centre and State. The Scheme is merged with another Scheme named SGSY since 1st April,1999.

Training of Rural Youth for Self-Employment (TRYSEM): The training of Rural Youth for self-employment (TRYSEM) is a supporting component of the IRDP, started as a centrally sponsored scheme on 15th August, 1979. It aims at providing technical and entrepreneurial skills to rural un-employed youths in the age group of 18-35 years from the families below the poverty line to enable them to take up income generating schemes. This scheme is no more in operation. It is merged with SGSY since 1st April, 1999.

Supply of Improved Tool-kits to Rural Artisans (SITRA): The programme is implemented as a part of IRDP. At the district level, the DRDA is the nodal agency. The scheme is formulated and circulated to all the State Governments on 20th July, 1992. Under this programme any suitable improved hand tool is to be provided. All the prudential rural artisans will be able to enhance the quality of the product to increase their production and their income and lead a better quality of life. No more in operation, it is merged with SGSY since 1st April, 1999.

Development of Women and Children in Rural Areas (DWCRA): The Development of Women and Children in rural areas (DWCRA) programme was launched as a sub-component of IRDP and a centrally sponsored scheme of the Department of Rural Development with UNICEF cooperation to strengthen the women's component of poverty alleviation programmes. It is directed at raising the income levels of women of poor households so as to enable their organised participation in social develeopoment towards economic self-reliance. The DWCRA's primary thrust is on the formation of groups of 15 to 20 women from poor households at the village level for delivery of services like credit and skill training, cash and infrastructural support for self-employment. Through the strategy of group formation, the programme aims to improve women's access to basic services of health, education, child care, nutrition and sanitation. It is merged with SGSY since 1st April, 1999.

Ganga Kalyan Yojana (GKY): It is a centrally sponsored scheme, being launched with effect from 1st February, 1997. The objective of the scheme is to provide irrigation through exploitation of ground water (borewells and tubewells) for individual and group of beneficiaries belonging to the target group. This scheme is no more in operation. It is merged with SGSY since 1st April, 1999.

Million Wells Scheme (MWS): Million Wells Scheme was taken up as a sub-scheme of NREP (National Rural Employment Programme) and Rural Landless Employment Guarantee Programme (RLEGP) during the year 1988-89 has continued under JRY. Till 1989-90, the objective of the scheme was to provide open irrigation wells to small and marginal farmers amongst the Scheduled Castes/ Scheduled Tribes and freed Bonded Labourers who are below poverty line, free of cost. From 1990-91 onwards, under Million Wells Scheme, the following works were also included: (a) Construction of open irrigation wells for the target group; (b) Where wells are not feasible, other schemes of minor irrigation like irrigation tanks, Water Harvesting Structures for the benefit of target group can also be taken up; (c) The provision under MWS can also be utilised for the land development of the target groups. From 1st January, 1996, this scheme had been delinked from JRY and made an independent scheme by itself. Now the scheme is no more in operation and merged with SGSY since 1st April, 1999.

Indira Awas Yojana (IAY): Indira Awas Yojana (IAY) which was launched during 1985-86 as a sub-scheme of RLEGP has continued as part of JRY since its launch on April 1989. However from 1st January, 1996, IAY has been made a separate scheme. The objective of IAY then was to provide dwelling units, free of cost to the members of Scheduled Castes/Scheduled Tribes and freed Bonded Labourers living below the poverty line. From 1993-94, the scheme has been extended to non-SC/ST rural poor also. Indira Awas Yojana is a centrally sponsored scheme funded on cost sharing basis between the Government of India and the State Government in the ratio of 75:25. The cost of IAY houses have been enhanced from Rs. 14,000 to Rs. 20,000 in hilly and difficult areas.

Jawahar Rojgar Yojana (JRY): Alleviation of rural poverty has been one of the main objectives of the development programmes. Since independence various schemes of employment generation were taken up from time to time in the country. The Eighth Plan has also stressed the need for having a larger focus on the programmes aimed at giving self-employment and wage employment to the poorer section of the community. During the first four years of the Seventh Five Year Plan, two Wage-employment Programmes, viz. NREP (National Rural Employment Programme) and Rural Landless Employment Guarantee Programme (RLEGP) were in operation in the country. From 1st April, 1989, i.e. last year of the Seventh Five Year Plan, these programmes were merged into a single wage employment programme known as Jawahar Rojagar Yojana (JRY). The primary objectives of JRY are generation of additional gainful employment for the un-employed and under-employed men and women in rural areas. The secondary objectives of this programme are creation of sustainable employment by strengthening the rural economic infrastructure.

Intensified JRY (IJRY): The second stream of JRY called Intensifed JRY has been introduced from December 1993 in 2 districts of the State. Phulbani district is one of them. The objective of the scheme is to intensify the efforts for rural employment. All employment works resulting in creation of durable productive community assets providing employment on sustained basis may be taken up under the scheme. The basket of schemes may include construction of all weather road, minor irrigation works, soil and water conservation works, water harvesting structures, watershed development, farm forestry etc. Works for strengthening rural infrastructure like primary schools, markets in especially different tribal areas with appropriate supplementary funds from other sources are also taken up. Wage and non-wage ratio under the work should be 60:40. Now it is working in the field of spice development.

Operation Black Board (OBB): Universalisation of primary education is the basic input for acquisition of functional skills which are absolutely essential for promoting self-reliance of the rural poor and their children. Primary education pre-supposes a basic infrastructure, i.e. a school building with a library, facilities for drinking water, toilet etc. A large number of primary schools in the state does not have buildings. To fill up this gap construction of primary school buildings was taken up under OBB programme from the year 1990-91. Sixty per cent of funds for the scheme is met by additional JRY grant and 40 per cent from the grant of State School and Mass Education Department.

Employement Assurance Scheme (EAS): The Employment Assurance Scheme (EAS) aims at providing wage employment in unskilled mannual works to the rural poor who are in need of employment and seeking it. The secondary objective is to create economic infrastructure and community assets for sustained employment and development. The Employment Assurance Scheme for generating employment opportunities to the rural poor on an assured basis has been launched from 2nd October, 1993. The scheme is the single wage employment programme implemented at the district/ block level through out the country. The scheme is operative in all the 12 Blocks of this district. A maximum of two adults per family are provided 100 days employment on an assured basis, which need and seek wage employment during the lean agriculture season. The ressources under the scheme would be shared between the Centre and the State in the ratio of 75:25 respectively. Men and women above 18 years of age and below 60 years of age normally residing in the village are covered.

Krushak Kalyan Karyakarma (KKK): Rs. 45.00 lakhs were received from Government for the execution of Borewells in this district. Since it is not feasible to construct Borewell as per Geological Survey Report, Government was moved to allow for construction of River Lift Irrigation under the scheme. Subsequently it was decided in the Collecrtor's Conference that un-utilised funds under KKK will be utilised for construction of L.I. Points/Diversion Weirs. Accordingly 2 L.I.Ps for Rs. 6.24 lakhs and 21 Diversion Weirs for Rs. 39.60 lakhs making the total of Rs. 45.84 lakhs have been taken up by the D.M., OAIC. and ASCOs respecrtively. The excess amount of Rs. 0.84 lakhs sanctioned for execution of Diversion Weirs was adjusted out of EAS funds. This scheme is no more in operation.

Rural Connectivity Programme (RCP): The broad objectives of the Rural Connectivity Programme in the State is to provide all weather connectivity as per the prescribed specification from the district to Sub-divisional headquarters, Block headquarters to Tehsil headquarters and from Panchayat Samiti headquarters to Gram Panchayat headquarters in order of priority, in original Guidelines of Rural Connectivity Programme (RCP). RCP funds received for connectivity under Tenth Finance Commission Award (TFC) will be utilised for construction of all weather roads as per action plan approved by the Zilla Parishad. The scheme is implemented since 1996-97.

Swarnajayanti Grama Swarojgar Yojana (SGSY): The objective of Swarnajayanti Grama Swarojagar Yojana (SGSY) is to provide sustainable income to the rural poor. The programme aims at establishing a large number of Micro-enterprises in the rural areas building upon the potential of the rural poor. It is envisaged that every family assisted under SGSY will be brought above the poverty line in a period of three years. This scheme is launched on 1st April, 1999, the programme replaces the earliar Self Employment and allied programmes IRDP, TRYSEM, DWCRA, SITRA, GKY and MWS, which are no longer in operation. The programme covers families under below poverty line in rural areas of the country within this traget group, special safeguard have been provided by reserving 50 per cent of benefits for SC/STs, 40 per cent for women and 3 per cent for physically handicapped persons subject to availability of funds. It is proposed to cover 30 per cent of the rural poor in each block in the next five year. SGSY is a credit-cum-subsidy programme. It covers all aspects of self-employment such as organisation of the poor into self-help groups training, credit technology, infrastructure and marketing. SGSY is a centrally sponsored scheme and funding shared by the Central and State Governments in the ratio of 75:25.

Jawahar Gram Samridhi Yojana (JGSY): Jawahar Gram Samridhi Yojana (JGSY) is the restructured streamlined and comprehensive version of erstwhile Jawahar Rojagar Yojana, designed to improve the quality of life of the poor, JGSY has been launched on 1st April, 1999. The primary objectives of the JGSY is creation of demand driven community village infrastructure including durable assets at the village level and assets to enable the rural poor to increase the opportunity for sustained employment. The secondary objective is the generation of supplementary employment for the unemployed poor in the rural areas. The wage employment under the programme shall be given to Below Poverty Line (BPL) families. JGSY is being implemented entirely at the village Panchayat level. Village Panchayat is the sole authority for preparation of the Annual Action Plan and its implementation. The programme will be implemented entirely as a centrally sponsored scheme on cost sharing basis between the Centre and the State Government in the ratio of 75:25.

Drought Prone Area Programme (DPAP): The Drought Prone Area Programme (DPAP) aims to mitigate the adverse effect of drought on the production of crops and livestock, productivity of land, water and human resources. It strives to encourage restoration of ecological balance and seeks to improve the economic and social condition of the poor and the disadvantaged sections of the rural community. Now DPAP is a people's programme with Government assistance. There is a specific arrangement for maintenance of assets and social audit by Panchayati Raj Institutions. Development of all catagories of land belonging to Gram Panchayat, Government and individuals fall within the limits of the selected watersheds for development. Allocation is to be shared equally by the Centre and

State Governments on 50:50 basis Watershed Committees is to contribute for maintenance of the assets created. Utilisation of 50 per cent of allocation under the Employment Assurance Scheme (EAS) is for the Watershed Development funds are directly released for sanction of projects and release of funds to Watershed Committees and Project Implementing Agencies (PIAs). Village community including self help groups undertake area development by planning and implementation of projects on watershed basis through Watershed Associations and Watershed Committees constituted from among themselves. The Government supplements their work by creating social awareness imparting trainings and providing technical support through the Project Implementation Agencies.

Self Help Groups

The Self Help Groups (SHGs) Guiding Principle stresses on organizing the rural poor into small groups through a process of social mobilization, training and providing bank credit and government subsidy. The SHGs are to be drawn from the BPL list approved by the Gram Sabha wherein about ten persons are selected, one each from a family and focus on the skill development trainings based on the local requirement. The SHG movement has gathered pace in countryside and is directly or indirectly contributing towards the economic development of rural areas. This article throws light on forming self help groups in rural areas as well as its effect on the rural economy.

Today, over 2,90,500 women from some of the most deprived areas of Uttar Pradesh can proudly claim to have engineered a socio-economic revolution that has impacted the lives of thousands of rural poor. These agents of change are members of Community Institutions of the Poor (Self Help Groups and their federations), driven and managed by the poor themselves which have helped disadvantaged communities unleash their potential for progress. The catalyst in this powerful transformation has been the Rajiv Gandhi Mahila Vikas Pariyojana (RGMVP), the Rajiv Gandhi Charitable Trust's flagship programme for rural development. Since 2002, when it was initiated, RGMVP has worked relentlessly to build a scalable institutional framework that continues to give a voice and freedom from social and economic oppression to a growing number of poor while serving as a conduit for them to gain access to a wide range of development initiatives.

This objective is to be achieved by organizing the rural poor into SHGs through a process of social mobilization, their training and capacity building along with the provision of income-generating assets through a mix of bank credit and government subsidy. The main objective of SHG concept is to improve the economic development of women and create facilitating environment for their social transformation in the lift of gender discrimination in work and the household.

MNREGA

The act directs state governments to implement MNREGA "schemes". Under the MNREGA the Central Government meets the cost towards the payment of wage, 3/4 of material cost and some percentage of administrative cost. State Governments meet the cost of unemployment allowance, 1/4 of material cost and administrative cost of State Council. Since the State Governments pay the unemployment allowance, they are heavily incentivized to offer employment to workers.

However, it is up to the State Government to decide the amount of unemployment allowance, subject to the stipulation that it not be less than 1/4th of the minimum wage for the first 30 days, and not less than 1/2 of the minimum wage thereafter. 100 days of employment (or unemployment allowance) per household must be provided to able and willing workers every financial year.

Provisions under MNREGA

Adult members of a rural household, willing to do unskilled manual work, may apply for registration in writing or orally to the local Gram Panchayat.

(i) The Gram Panchayat after due verification will issue a Job Card. The Job Card will bear the photograph of all adult members of the household willing to work under MNREGA and is free of cost.

(ii) The Job Card should be issued within 15 days of application.

(iii) A Job Card holder may submit a written application for employment to the Gram Panchayat, stating the time and duration for which work is sought. The minimum days of employment have to be at least fourteen.

(iv) The Gram Panchayat will issue a dated receipt of the written application for employment, against which the guarantee of providing employment within 15 days operates.

(v) Employment will be given within 15 days of application for work, if it is not then daily unemployment allowance as per the Act, has to be paid, liability of payment of unemployment allowance is of the States.

(vi) Work should ordinarily be provided within 5 km radius of the village. In case work is provided beyond 5 km, extra wages of 10 per cent are payable to meet additional transportation and living expenses.

(vii) Wages are to be paid according to the Minimum Wages Act, 1948 for agricultural labourers in the State, unless the Centre notices a wage rate which will not be less than ₹60 per day. Equal wages will be provided to both men and women.

Note: The original version of the Act was passed with Rs. 60/day as the minimum wage that needs to be paid under MNREGA. However, a lot of states in India already have wage regulations with minimum wages set at more than ₹100 per day.

MNREGA's minimum wage has since been changed to ₹100 per day. (Against the Rs. 100 per day wages originally approved, actual wages given are as high as Rs. 181 in Nicobar. Haryana gives Rs. 179 per day while daily wage in Chandigarh under the scheme is Rs. 174. It is Rs. 170 in Andaman and Rs. 150 in Kerala. The wages are in the range of Rs. 118 to Rs. 138 per day in other states).

(viii) Wages are to be paid according to piece rate or daily rate. Disbursement of wages has to be done on weekly basis and not beyond a fortnight in any case.

(ix) At least one-third beneficiaries shall be women who have registered and requested work under the scheme.

(x) Work site facilities such as crèche, drinking water, shade have to be provided.

(xi) The shelf of projects for a village will be recommended by the gram sabha and approved by the zilla panchayat.

(xii) At least 50 per cent of works will be allotted to Gram Panchayats for execution.

(xiii) Permissible works predominantly include water and soil conservation, aforestation and land development works.

(ivx) A 60:40 wage and material ratio has to be maintained. No contractors and machinery is allowed.

(xv) The Central Government bears the 100 per cent wage cost of unskilled manual labour and 75 per cent of the material cost including the wages of skilled and semi skilled workers.

(xvi) Social Audit has to be done by the Gram Sabha.

(xvii) Grievance redressal mechanisms have to be put in place for ensuring a responsive implementation process.

(xviii) All accounts and records relating to the Scheme should be available for public scrutiny.

REFERERNCES

A. Rudra, "Organisation of Agriculture for Rural Development", in D. Ghai *et al.*, *Agrarian System for Rural Development*, Geneva, ILO, 1979.

Dr. Vivek Kumar Singh 'Reader', *Department of Social Work and Sociology*, Dr. Shakuntla Misra Rehabilitation University, Mohan Road, Lucknow.

Govermment of India, Five Year Plan, Planning Commission, New Delhi.

Lalit K. Sen, *et al., Planning Rural Growth Centres for Integrated Area Development*, Institue of Community Development, Hyderabad, 1971.

M.N. Srinivas, "On Living in a Revolution", *Vierteljahresberichte*, Vol. 110, 1987. Srinivas provides an illuminating analysis on this.

R.P. Mishra: Growth Poles and Growth Centres in the Context of India's Urban and Regional Devlopment Problems in *Growth Poles and Growth Centres Regional Planning*, (Ed.) Antoni R. Kuklinski, United Nations Research Institute for Social Development, Geneva.

S.R. Maheshwari, *Rural Development in India*, New Delhi, Sage, 1985. He provides another typology and policy descriptions.

V. Dhagamvar and N. Rath, "Poverty in India", paper presented to Workshop of Poverty in India, Oxford University, 1987. He documents the poor design and ineffectiveness of Labour Laws.

V.K.R.V. Rao, "Integrated Rural Development"— Paper presented to the Third Biennial Conference of Association of Development Research and Training Institutes of Asia and the Pacific at Goa, 1977.

V.K.R.V. Rao: *op. cit.*

V.M. Rao: *op. cit.*

VLS Pradasa Rao: *Development Strategy for an Agricultural Region*, Institute of Development Studies, University of Mysore, 1976.

Waheeduddin Khan, *et al., Plan for Integrated Rural Development, in Pauri Gharwal*, National Institute of Community Development Hyderabad, 1976.

18

Need Based Village Development Planning

H.R. Yadav

1. Introduction

The planning is an organized process by which the people achieve its development of goals. If the intentions and objectives of development are distorted them the planning also becomes a distorted exercise. Planning is done without willing consent of the local problem and without involvement of the people and their participation because the planning is talk of the people in which planners can provide techno-scientific backup support to achieve their planning goals. Village development is the new system of decentralization. Local people should be done by the local people. Faster development will be ensured assuming village as a centre for development. Real administration is transferred to the social village level bodies in which the planning development works, if management and control is ensured through peoples participation.

Planning should be done from bottom to top. Revolutionary step of decentralization will have complete change among the villagers' life and it will have new era of integrated rural development of the rural areas at grassroots level. The decentralization of administrative and financial powers of village Panchayats have been ensured by the government. In this study an attempt has been made to analyze the need based village development planning as per the requirement of down-trodden deprived rural poor people. The most important amenities, viz. health, sanitation, education, drinking water, food, welfare programmes and various other schemes being implemented and supported by the government has been analyzed and need based planning process has been formulated to fulfil the ever increasing demand of the local village people.

In view of the above "Self Government" of Panchayat and need based people participatory action plan has been formulated to fulfil the dreams of "Self Government", i.e. Panchayat Raj to Lok Raj to Gram Raj to establish the Ram Raj in the village which has been the vision and dream of Mahatma Gandhi, so that such deprived rural people of the society are able to more equally with the upper section of the society as constitutional rights of our nation.

2. Objectives

The main objectives of the study are to analyze the need based action oriented Village Development Planning of the Village Panchayat Sarai Khema of Amethi Block.

3. Need Based Village Planning

Technocrats/Scientists and planners should assist villagers as a social worker who can easily understand the basic grassroots level needs of the villagers, so that needs of the people are properly planned and rural development activities are implemented with the ensured participation of village people to achieve the planning target of integrated development.

The Village Panchayats have been given requisite responsibilities, administrative control, physical, financial powers and concerned implementing line departments have also been handed over to the Village Panchayat to ensure Self-Government of the Village.

In view of the "Self-Government" of village Panchayat Sarai Khema need based people's participatory action plan has been formulated to fulfil the dreams of "Self-Government", i.e. "Panchayat Raj to Lok Raj to Gram Raj to Ram Raj" which has been the vision and dream of Mahatma Gandhi Ji.

In view of the above various administrative and financial rights have been allocated to the village Panchayat under the decentralization system of Panchayati Raj Institutions. It has been observed that Panchayati Raj Institutions have been strengthened and made more autonomous for the formulation of need based planning, implementation and maintenance of the planned programmes, various developmental works have been handed over to the village Panchayat for implemented programmes.

Village Panchayats have been provided the work responsibilities 'education' in which primary, junior, senior secondary schools, non-formal education programmes are transferred to village Panchayats, school teachers, Head Masters, instructors are working in the control of village Panchayats for which requisite financial and administrative supports and rights have been provided to the village Panchayats.

Government "Tube-wells" are also handed over to the village Panchayat and its maintenance/management agreement has been ensured with the minor irrigation department.

Drinking water—India Mark-II Hand Pumps are the assets of village Panchayat, for repair, maintenance of the hand pumps, as per the norms, the requisite finance has been provided by the Government directly to the village Panchayat.

Youth, Welfare and Support Programmes are being implemented by the Village Panchayats for which financial support and materials have been provided by the Government to the village Panchayats.

Medical, Health and Family Welfare Programme are being implemented under the control of village Panchayats.

Village Level Women and Child Development Programme are implemented under the control of Village Panchayats.

Animal Service Centre and 'D' Category Animal Hospital are controlled and managed by the Village Panchayats.

Food Supply Shops, Distribution and Allotment Rights are given to the Village Panchayats.

Village Level Agricultural Development Programmes are implemented by the Village Panchayats.

Village Level Rural Development Programmes are implemented, maintained and managed by the Village Panchayats for which financial support has been provided by the Government.

Village Level Related Works of Panchayati Raj Institutions are implemented and controlled by the Village Panchayats.

Village Panchayats are also distributing **Pension, Scholarship** and other such facilities in the open meeting of the Village Panchayats.

Need based village level planning for the village Sarai Khema has been planned. It has been observed that at village level most of the works are being implemented for the construction of Nali, Kharanza, Chackroad, Pulia repair and maintenance of the community assets for which maximum financial support to village Panchayats are made available.

Need based action plan is approved by the Village Panchayat in which planning, execution, monitoring and supervision are also considered. Monitoring Committee is formed involving Scheduled Caste and Women as member.

The major need based people's participatory action plan of the Village Panchayat Sarai Khema has been mainly the construction of Nali, Kharanza, Chak roads, earth work, construction of ponds and maintenance of the community assets, besides there are several other need based programmes which can increase the agricultural productivity, to generate more employment and increase the income of the deprived rural multitudes through optimum utilization of available natural and human infrastructure in the village Sarai Khema. In view of the above, an attempt has been made to analyse the need based people's participatory action plan of village Sarai Khema of the Amethi Block, District Amethi.

1. *Repair and Maintenance Work:* Village Panchayat Sarai Khema resolved to repair, maintain, white washing/colouring of the Panchayat Bhawan and purchase of office stationary materials etc.
2. *Hand Pump:* Village Panchayat Sarai Khema also passed resolution for the maintenance of India Mark II Hand Pumps and construction of the platform for installed India Mark II Hand Pumps in the village Sarai Khema and its revenue villages and hamlet villages.
3. *Light Facility:* It has been resolved by the Village Panchayat Sarai Khema for the purchase of C.F.L. Sodium Light for facilitating light in the village streets.
4. *Water Drainage:* Village Panchayat Sarai Khema has also resolved to purchase Hume Pipe for the water drainage to drainout the water logging problem of the village.
5. *Repair of the Sanitation:* Village Panchayat Sarai Khema resolved to repair the sanitation facility, i.e. repair and maintenance of latrines constructed at the Primary Schools and also repair of the latrine constructed at the Anganwadi Centres in village Sarai Khema of the Amethi Block.
6. *Remuneration:* It has been resolved by the Village Panchayat Sarai Khema for the remuneration payment of the Village Panchayat Sarai Khema.

7 *Annual work plan of the XIII Finance Commission of the State Finance:* Village Panchayat Sarai Khema passed resolution for the annual work plan of the State Finance under 13th Finance Commission is as follows:-

(i) Repair and maintenance of constructed kharanja road.

(ii) Repair and maintenance of kharanja near Budha Mai place.

(iii) Repair and maintenance of kharanja from Mata Kali temple to the residence.

(iv) Repair and maintenance work of the constructed kharanja road from the residence of Brijlal to the residence of Ram Pher Kashap in village Sarai Khema of the Amethi Development Block.

It has been observed that the Village Panchayat has proposed the construction, repair and maintenance of the various programmes in the village Sarai Khema within the limited allocated financial support provided by the State Government. Besides this there are various other welfare programmes which are supported by the Block and District Panchayat, District Rural Development Agency, State Government and Government of India. Integrated village development need based planning should be prepared by the Village Panchayat which should be supported by the various government agencies in the interest of the villages with the view to generate employment and increase the income of deprived rural poor surviving under Below Poverty Line (BPL).

4. Need Base Integrated Village Development Action Plan

Need base integrated village development action plan for village Sarai Khema is as follows:

1. **Land Development**: Land development including reclamation of 25 hectares uncultivable land and non-agricultural wasteland 90 hectares and 22 hectares other uncultivated land and 15 hectares old fallow land lying waste from 2-3 years the available banjar land has to be reclaimed for agriculture, horticulture and social forestry programmes to meet the increasing demand of Fuel wood, Fodder, Fibre, Food grains, Fruits and Vegetables. The available 39 hectares

waterlogged land has to be reclaimed for developing Ponds for Fisheries and Social Forestry and Agro-Forestry Programmes to generate employment and increase the income of deprived rural multitudes.

2. **Scheduled Caste Patta Waste Land**: Excess available waste lands have been allotted on Patta to the landless scheduled caste beneficiaries of the village. Such Patta land allotted scheduled caste beneficiaries are the owners of such land and the land has been entered in the land revenue records for which such land holders are liable to pay the Malgujari as per the revenue norms. Thus such patta allotted waste lands should be reclaimed under MNREGA schemes for Agri-Horti-Afforestation programmes to conserve the environment, generate employment and increase the income of deprived rural poor of village Sarai Khema and to meet the increasing demand of Fuel wood, Fodder, Fibre, Fruits, Food grains, Fisheries and Vegetables for the growing cattle and human population.
3. **Small Scale and Agro-Industries**: Small scale and agro-industries has to be promoted for the value addition of agricultural products and to generate employment and increase the income of deprived rural multitudes of the village Sarai Khema.
4. **Drinking Water:** India Mark II Hand Pumps should be installed as per the needs of the local people not on the basis of vested interest of political people's representatives.
5. **Rural Road:** Rural roads including Pacca road, Kuccha road, Kharanja, Chakroad, pulia etc. should be constructed as per the local needs of the village people.
6. **Energy**: Electrification and non-conventional sources of energy including solar light etc. should be installed and facilitated as per needs of the local village people not on the basis of voting interest of the Village Panchayat representatives.
7. **Welfare Programmes**: Scheduled Caste, Backward caste, Minorities and Below Poverty Line Welfare Programmes should be provided as per the needs of the people of the Village Sarai Khema not on the basis of the vested interest of the people's representatives.

8. **Medical and Health**: All the schemes related to medical and health including programmes of National Rural Health Mission (NRHM) should be properly implemented and facilitated to the needy interested people of the village Sarai Khema without any vested interest.
9. **MNREGA**: Mahatma Gandhi National Rural Employment Guarantee Act (MNREGA) schemes should be properly implemented and needy, down-trodden poor people should be given the MNREGA Job card. MNREGA schemes should also be implemented in the patta wasteland allotted to the scheduled castes through generating employment for the reclamation of their own land and to meet the increasing demand of Fuel wood, Fodder, Fibre, Fodder, Food grains, Fruits, vegetables and fisheries and also to conserve the environmental conditions of the village Sarai Khema.

5. Conclusion

People's participatory need based action plan for the village Sarai Khema of the Amethi Development Block of the District Amethi has been formulated and planned for the integrated village development to strengthen the "Self-Government" which has been the vision and dreams of the Mahatma Gandhi Ji various programmes have been handed over to the village Panchayat powers have also been handed over to the Village Panchayat to strengthen the Panchayat Raj Institutions.

Repair and maintenance of the Panchayat Bhawan, repair and maintenance of the installed India Mark II hand pumps installation of Sodium Light in the streets of the village, construction of water drainage, and repair and maintenance of the latrines constructed in the primary schools and Anganwadi centres and remuneration facility to the Village Pradhan etc. has been resolved as annual action plan for the implementation of village development programme in the village Sarai Khema.

Annual work plan of the State Finance under 13th Finance Commission has been repair, maintenance and construction of the Kharanja Road in the village Sarai Khema as per the needs of the local village people and people's representatives of the village.

Various other programmes like small scale and agro-industries, rural food, drinking water, energy, welfare programmes, medical and 'health', MNREGA and reclamation of Patta allotted waste lands to the scheduled caste beneficiaries for Agri-Horti-Afforestation to meet the increasing demand of growing cattle and human population for Fuel wood, Fodder, Fibre, Fodder, Food grains, Fruits, Fisheries and vegetables and also the conserve the environmental conditions of the village as well as to generate employment and increase the income of deprived rural poor multitudes of the village Sarai Khema of the Amethi Block.

19

Communication Strategy for Sustainable Development: The Indian Experiment

Mahendra Kumar Padhy

Introduction

The concept of sustainable development has occupied a central place in every aspect of human life today. It is a multidimensional and multidisciplinary concept covering almost all spheres of human activity. Sustainable development has become the concern of economists, ecologists, administrators, lawyers, communication, exports, environmentalists, human right activists, feminists, scientists and NGOs. In other words, it has become everybody's cup of tea. Since the present study aims at studying the role of communication in sustainable development without identifying the various implications of the concept, therefore, an attempt has been made in this research paper to discuss the various implications of sustainable development.

> The World Commission on environment and Development (WCED, 1987) defined sustainable development as the 'development that meets the needs of the present without compromising the ability of future generations to meet their own needs.'

It is observed that sustainable development is a coin which consists of two obligations on its two sides. One side is the alleviation of poverty and the other, the protection of environment. Sustainable development is very much linked with the involvement and active participation of people. It is a holistic concept that can be on the global, national, local and individual scale. Communication is an intervening variable without which the materialization of different goals of sustainable development is not possible. Therefore, Communication has the key role in facilitating the participation of people relating to sustainable development.

Conceptual Framework

The present study has been taken up with the following theoretical framework. There are a myriad of theories and models of communication, but there are only a few theories and models which deal with the questions of development. Therefore, it is useful to discuss the relevant models and theories in the context of the present study as follows.

Development media theory deals with the task of media in developing countries. It emphasizes the positive uses of the media in national development and for the autonomy and society. To a certain extent elements of this theory favour democratic and grassroots involvement, thus promoting participative communication models. (Mcquail, 1987). The one thing of the media is the acceptance of economic development itself and often the correlated nation building, as an overriding objective. To this end, certain freedom of the media and of journalists is subordinated to their responsibility of helping in this purpose. Collective ends rather than individual freedoms are emphasized. With the failure of the Dominant Paradigm of development, and its communication approach in bringing about the expected change, there took place a thinking about the alternative paradigm of development which led to the emergence of the concept of another development and subsequently a more specific one, sustainable development. With regard to communication also, a major shifts has taken place from top-down authoritative model of communication to a two-way horizontal and participatory model of communication.

Conservation of Biodiversity

Conservation of biodiversity is an indispensable component of sustainable development. The human survival and well-being depends upon the entire life sustaining system. The biosphere constitutes a vital life support system for man. Its existence in a healthy and functional state is essential for the existence of human race. It is the complex collection of innumerable organisms (including seas, forests and sky birds) and it is the biodiversity which makes our lives both pleasant and possible. Scientists believe that the total number of species on earth is between 10 million and 80 million. We have been able to enlist only 1.4 million species so far. Nature has taken more than 600 million years to develop this exceedingly complex spectrum of life on this planet. The existence of human race depends on health and well-being of other life forms in the biosphere. We are losing these accumulated heritages of millions of years at a very fast rate. An estimated 7.5 million hectares of tropical forests are cut down each year in developing countries and this pace is increasing. Humans are causing the extinction of perhaps one species an hour. It is estimated that about 25 per cent of the world's plant species might be lost by the year 2025 AD if the current rate of plant genetic erosion continues.

The erosion of biodiversity has serious repercussions on the health of the eco-systems on which we all depend, causing widespread impoverishment, as rural populations lose their sources of livelihood, food, and medicines as three-fourths of the worlds populations still uses plants and animals for medicinal purposes and eroding the genetic base of agriculture. Preservation of biodiversity is essential for a successful food security and breeding programme. If we ignore plant genetic erosion, it may lead to losing sources of resistance to pests, diseases and climatic stress, leading eventually to crop failure and famine. In fact, human survival itself is imperilled due to biodiversity loss.

Recognizing the need of conserving biodiversity, the Convention of Biological Diversity was signed at the Earth Summit in June 1992. By now, the CBD has been ratified by over 160 countries. The Biological Diversity Convention is fundamentally based on promoting the sustainable use of the components of biological diversity in a manner at a rate that does not lead to the long term decline of biological diversity, thereby meeting its potential to meet the needs of present and future generations.

Thus, the conservation of biological diversity is an important implications of sustainable development for communication. Various media of communications have a potential role to play in conserving biological diversity. The conservation of biological diversity is not possible unless the people are aware of the importance of conserving biological diversity. Therefore, in creating awareness and in conscientizing people about the sustainable development, various channels of communication are of crucial importance.

Significance of this Research Study

There have been many studies carried out on development and communication, media and development, environment and media, environment and communication and communication, rural development and communication media, traditional folk media and development and participatory development communication. But, though sustainable development is the latest and present trend of development, so far, proper attention has not been paid to this area from communication point of view. Therefore, it has been felt worthwhile to study the role of communication for sustainable rural development.

Research Questions

Sustainable development being the latest and the present trend of development, the broad aim of the study is to analyze

- the role of communication in sustainable development, and
- to recommend a suitable communication strategy for sustainable development.

Objectives of this Study

1. To identify the implications of sustainable development.
2. To find out the policies and programmes of Indian Government towards sustainable development.
3. To study and analyze the role of communication science in sustainable development.

An empirical study has been carried out in a backward and environmentally affected district in Orissa.

The following aspects were sought to be examined in Koraput, Orissa:

(a) Role of Communication in the success or failure of poverty alleviation programmes.
(b) Awareness about environment, Family Planning and reflection of that awareness in their lifestyles.
(c) Source of information to people and their media habits.
(d) Role played by various communication media in relation to combating pollution and in connection with environmental movements.
(e) To recommend a suitable communication strategy for sustainable development.

Role of Communication for Sustainable Development

Communication is a basic instinct of man. It is the fact of life of not only human beings, but also of animals, birds and other living beings. Communication maintains and animates life. It is also the expression of social activity and civilization. It leads people from instincts to inspiration through various processes and systems of enquiry, command and control. Communication integrates knowledge, organizations and power and runs a thread linking the earliest memory of man to his noblest aspiration through constant thriving for a better life. As the world has advanced, the task of communication has become ever more complex and subtle to liberate mankind from want, oppression and four and to write it in community and communion, solidarity and understanding. Mass communication comprise the institutions and technology by which specialized groups employ technological devices (press, radio, films etc.) to disseminate symbolic content to large, heterogeneous and widely dispersed audiences.

Poverty eradication, protecting the environment, reducing the consumption of non-renewable resources and increasing the use of renewable resources, conservation of biological diversity, controlling various types of pollution, land degradation and deforestation, waste management using appropriate technologies land reforms, population control and stabilization, upholding basic human rights, social welfare and women's upliftment, promoting intra-generational and intergenerational equity and participation of people from individual, local levels to global level, being the various important objectives of sustainable development, different communication channels have a potential role to play in fulfilling these objectives. Though communication alone is not sufficient to meet these objectives. It is a crucial element in facilitating the fulfilment of these objectives.

Communication Policy and Strategy for Sustainable Development

Strategies that include communication for sustainable rural development as a significant aspect of agricultural and rural development are sorely needed. Efforts in this direction are being made, but governments have yet to recognize fully the potential of this factor in promoting public awareness and information on agricultural innovations, as well as on the planning and development of small business, not to mention employment opportunities and basic news about health, education and other factors of concern to rural populations, particularly those seeking to improve their livelihoods and thereby enhance the quality of their lives.

Rural development is often discussed together with agricultural development and agricultural extension. In fact "agricultural extension" is often termed "rural extension" in the literature. In contrast, rural development includes but nonetheless expands beyond the confines of agriculture, and furthermore requires and also involves developments other than agriculture. Accordingly, government should consider the establishment of a communication policy that while supporting agricultural extension for rural development also assumes the role of a "rural extension" service aimed as well at diffusing non-agricultural information and advice to people in rural areas. A communication policy would aim to systematically promote rural communication activities, especially interactive radio but also other successful media such as tape recorder and video instructional programmes. Computers and the Internet may not yet be accessible to rural communities but they serve the communication intermediaries and agricultural extension agents who provide information to rural populations. Other devices such as cell phones hold considerable promise for the transfer and exchange of practical information. For reaching the final agricultural and basic needs information users in rural areas today, radio is the most powerful and cost-effective medium.

However, other traditional and modern communication methods are equally valuable, depending on the situation and availability, like face-to-face exchanges (via demonstration and village meetings); one-way print media (such as, newspapers, newsletters, magazines, journals, posters); one-way telecommunication media (including non-interactive radio, television, satellite, computer, cassette, video and loud-speakers mounted on cars); and two-way media: [telephone, including teleconferencing, and interactive (Internet) computer]. Information and communication technologies (ICTs) have proved to be important for Internet users and for the intermediate users who work with the poor. Pilot experiences show that various media are valuable for assisting agricultural producers with information and advice as to agricultural innovations, market prices, pest infestations and weather alerts.

ICTs also serve non-farming rural people with information and advice regarding business opportunities relating to food processing, wholesale outlets and other income-generating opportunities. In the case of non-agricultural rural development interests, a communication for rural development policy would aim to promote diffusion of information about non-agricultural micro-enterprise development, small business planning, nutrition, health and generally serve to provide useful, other than agriculture information.

By its very nature as mass media, communication for rural development can provide information useful to all segments of rural populations. However, it would serve as a first effort towards advancement of "rural extension" services and activities aimed at rural development concerns beyond those of agriculture. Thus, extension and communication activities would be expected to work in tandem, allied in the common cause of supporting income-generating activities, both agricultural and non-agricultural.

Concluding Remarks

The discussion mentioned above shows that though there are many definitions and multiple dimensions to sustainable development, these definitions and dimensions are not contradictory to each other but they corroborate each other. Broadly, the sustainable development can be described as the poverty alleviation, i.e. to enable the present generations to meet their needs and environmental protection to enable the future generations to meet their needs. In relation to communication, it implies that communication in general and various communication channels in particular have a vital role to play in creating awareness about the various poverty alleviation programmes initiated by the government; in the problem articulated by poor, and thus, in bridging the gap between the planner and the beneficiary. Environmental protection and promotion and population control being the other broad dimensions of sustainable development, various communication channels have a responsible role to play in informing, educating and conscientizing the people about various environmental issues and promotional programme and sustainable use of natural resources, using renewable sources of energy, conservation of biological diversity, waste management, prevention and control of pollution, family planning, etc. Besides, communication is of vital importance in promoting human rights, gender equality, social welfare and peace. Since the sustainable development calls for the participation of people—individuals, national and international levels, various communication channels are of great importance in facilitating the participation of people from individual, local levels to global level. Besides, since sustainable development calls for a lifestyle that is equitable and environment-friendly in moulding such lifestyle throughout the globe, communication has a potential role to play.

REFERENCES

Kumar, Keval J., *Mass Communication in India*, Bombay: Jaico Publishing House, 1987.

Desai, Vasant, *Rural Development*, Vol. 1. Himalaya Publishing House, 1988.

Kuppuswamy, B., *Communication and Social Development in India*, Bombay: Media Promoters and Publishers Pvt. Ltd., 2000.

McBride, Sean, UNESCO International Commission for the Study of Communication Problems, Many Voices One World, UNESCO, 1980.

Maheswari, Henry *et al.* (2005); *Learning from the Rural Poor*: Indian Social Institute, New Delhi.

Wang Georgette and Wimal Dissanayake, *The Study of Indigenous Communication Systems on Development: Phased out or Phasing in Media Asia*. Vol. 9, Number 1, 1982, p. 3.

20

Role of Community Radio in Rural Development-Based Communication

Ravi Suryavanshi and *Neha Nigam*

Introduction

India is a developing country which is striving hard for development. Rural areas, villages are the main concern in all dimensions of developmental issues. Government, NGOs and many agencies working on this aspect and many programmes also running like National Rural Health Mission, Asha, Anganbadi, National Rural Livelihood Mission, Rajeev Gandhi Awas Yojana, MNREGA, Aam Admi Bima Yojana and many more but the distress is most of the poor people are not conscious about all these ongoing projects. In this case Community Radio is becoming a strong medium that not only gives information and make alert but also entertain as well. Radio is the main source of news and entertainment for most of people in India. The basic attributes of radio are very cost effective, easily reached and even illiterates can use and get benefits of this all pervasive medium. Today community radio is playing a significant role in the rural development. Community Radio serves people living together and sharing same interests and geographical areas. Thus this service is able to appreciate and understand the specific needs and aspirations of locals including the rural population and formulate and produce programmes accordingly.

They have all opportunity for active participation in community and freedom of expression. With the help of community radio people can easily access the information that is needed by them and become more aware and alert. Community Radio follows, at least theoretically, democratic participation in mediation.

Community Radio in India

In India, in the mid 1990s, the campaign to legitimise community radio was started. And soon after in February 1995 Supreme Court of India declared in its judgment that "airwaves are public property". Since then it became source of inspiration for communities across the country. At that time only educational radio stations were allowed. Anna FM was India's first campus 'community' radio, launched on 1 February, 2004, which is being run by Education and Multimedia Research Centre (EMMRC). On 16 November, 2006, the government of India notified new Community Radio Guidelines which permit NGOs and other civil society organizations to own and operate community radio stations. The first community-based radio station, licensed to an NGO was launched on 15 October, 2008.

Andhra Sangham Radio in Pastapur village, Andhra Pradesh state was the first community radio is licensed to Deccan Development Society (DDS), an NGO that works with poor, rural, dalit women's groups in about 75 villages, with assistance from UNESCO in 1998. Some others are Radio Bundelkhand (Orchha, Madhya Pradesh), Mann Deshi Tarang (Satara, Maharashtra), Namma Dhwani (Budikote, Karnataka), radio Mattoli (Wayanad, Kerala), Kalanjiam Samuga Vanoli (Nagapattinam, Tamil Nadu) and Barefoot (Tilonia, Rajasthan) which are governed and run by NGOs. Government approved educational institutions, civil societies and voluntary organizations, State agricultural universities (SAUs), ICAR institutions, Krishi Vigyan Kendras, registered societies and autonomous bodies and public trusts registered under Societies Act, or any other such act relevant for the purpose can apply for Community Radio Station. Individuals, Political parties and their affiliate organizations (including students, women, trade unions and such other wings affiliated to these parties), organizations operating with a motive to earn profit, organizations expressly banned by the Union and State governments cannot apply for Community Radio. The Kutch Mahila Vikas Sangathan (KMVS), is also an NGO working with rural women in the villages of Kutch, Gujarat, known as Ujjas Radio. *Chala Ho Gaon Mein* is a community radio which has been broadcast from a backward region in the Palamau district of the state, Daltonganj.

Community radio is totally based and focused on issues like education, health, environment, and agriculture, rural and for community development. At least 50 per cent of content will be created with the participation of the local community, for which the station has been set up. Transmission of sponsored programmes will not be permitted except programmes sponsored by Central and State Governments and other organizations working in public interest. Local advertisements can be broadcast for 5 minutes every hour.

Community broadcasting contributes towards empowerment in that it trains individuals in broadcasting skills (DoCA, 1997); it forms local and extended networks (Melzer, 2000); and can provide 'a sense of local community identity' (Elson, 2000).

Community Radio station is aimed at covering an area of 5-10 km radius. Effective Radiated Power (ERP) specified is 100 W. However, in special cases, UPTO 250 Watt ERP can be considered subject to conditions laid down in the guidelines.

The Community Radio Forum, India, was registered as a 'Society' and 'Trust' on 26 February, 2008. The tower height to a maximum of 30m is allowed and that the minimum height is stipulated as 15m to prevent possibility of biological hazards of Radio Frequency (RF) radiation. Community radio stations are expected to produce at least 50 per cent of their programmes locally, as far as possible in the local language.

Scope of Community Radio

As says Aristotle, "A citizen is the one who plays the dual role of the ruler as well as the subject", i.e. a responsible citizen is the one who necessarily participates in the process of governance. Central to the community radio industry is a commitment to local access and participation to groups and individuals (Thomas, 1993). By 1 July, 2010, the Ministry of Information and Broadcasting, Government of India, announced that 715 applications for CR licences had been received. By 12 August, 2010, the number of operational community Radio stations in India had risen to 82. By December 2010, the number of operational community radio stations in India has risen to 103. The augmenting number of Community Radio stations shows that it has become very popular because of its features that cater to the interests of such groups and also makes them as active participants in core activities. This is enormously good for the health of the Indian democracy. Thus, community radio acts as a source of effective communication *vis-à-vis* various developmental works that are on and their implications for the local residents.

Conclusion

Community radio also known as rural radio, cooperative radio, or development radio as well. With the help of emerging new technologies India is getting a new jump into the overall development and the introduction of community radio is a good step towards the development of the people, which serves through rural development based communication.

In India, where literacy remains a considerable obstacle to development, community radio reach a large number of illiterate poor people and fulfil the purpose of giving them substantial information regarding their interest. In the more remote areas, Community Radio stations are almost the only media existing. In 2000, AIR programmes could be heard in two-third of all Indian households in 24 languages and 146 dialects, over some 120 million radio sets. Community Radio gives a voice to the community they serve with programmes in local languages, with regard to local culture, traditions and interests. Radio is a powerful communication tool. India's post-independence experiments with ICT use in agricultural development started with radio. Community radio faces the challenges in effective and excellence programme production in terms of content, production quality and community participation because of illiteracy and lack of awareness among people. Agriculture has always been an extremely knowledge-intensive sector requiring continuous information flow. The rapid changes happening around with globalization, uncontrolled, uncertainty in climate change, discerning consumer segment and continued farm crisis highlight the importance of timely, appropriate and need based information and knowledge to meet innumerable developmental challenges. Community Radio may help to empower rural or marginalized people and also inspire them with a feeling of self-confidence for redefining their up to now aimless lives. Thus the future of Community Radio regarding development of rural communities is very bright.

REFERENCES

Community Radio in India, Indian radio, www.indianetzone.com/37/communityradio India, 2010.

Community Radio Manual, www.gramvani.org/news/cr-in-india, 2010.

Elson, S. 2000 'Hello Community, this is John Laws: Feature', CBX, May, p. 13.

Fraser, Colin and Restrepo Estrada, Sonia, Community Radio Hand Book, UNESCO, 2001.

http://en.wikipedia.org/wiki/Community_radio

http://www.i4donline.net/news/news-details.asp?Title=India:-First-community-radio-of-MP-launchedandcatid=2andnewsid=15538

http://www.youthkiawaaz.com/2011/07/community-radio-as-an-effective-tool-for-agricultural-develcpment-research/

Melzer, D. 2000 'Free Radio: President's Report', CBX, May, p. 3.

Thomas, P. (1993), 'Alternative communications: problems and prospects', *Media Asia*, 20 (2), pp. 63-65. Department of Communication and the Arts 1997 Training in the Community Broadcasting Sector, DCA: Canberra.

Section IV
Rural Health Mission

Introduction

Rural Health is one of the basic programme, which has to be implemented at village level equally, effectively and efficiently for all. Rural Health Mission Programmes are not being properly implemented. The vision, mission, actions of the implementing institutions has been distorted and diverted from its main goal. The real beneficiaries at village level are not the real beneficiaries for whom National Rural Health Mission (NRHM) programmes are implemented. It has been realized that misappropriate planning and unplanned implementation of the programmes without peoples participation has been one of the major reasons which are responsible for such unwanted/undesired results.

In view of the above, H.R. Yadav, has attempted to analyze the planning for rural health mission programmes for the deprived rural poor multitudes. The problems and prospects of rural health including various programmes being implemented by the various agencies has been properly propagated and the interested rural poor beneficiaries have motivated for active participation. An effort has been made in this study to analyze, identify, ignored rural poor beneficiaries to make them benefited under the various programmes being implemented under the National Rural Health Mission Programme.

Manish Yadav and Uttara Yadav have tried to analyze the vehicular pollution and its health effect and its control. Major vehicular pollutants has been analyzed in this study and various measures taken for improvement in the quality of the automation fuel and the overall impact of these measurse and the future strategy to be adopted for vehicular emission reduction and related issues have been analyzed.

Rajneesh Chaudhary has made an effort to analyze the role of National Rural Health Mission (NRHM) in providing effective health care services to rural population in India. Various components of primary health system like preventive, promotive and curative care human resource management diagnostic services, logistic management, disease management and surveillance and improved service delivery etc. has been explained in this study.

Diwakar Yadav and Shashi Kant has explained the role of

Reformulations of the demographic transition perspectives have emphasized its revolutionary character and have shown that demographic transition is not one, monolithic change but rather than it encompasses several integrated transitions. A decline in fertility, and by subsequent transitions in migration, urbanization, the age structure and the family and household structure in society.

21

Planning for Rural Health Mission

H.R. Yadav

1. Introduction

It is known fact that the International Agencies, viz. UNICEF, WHO etc., Central Government-State Government-District-Tehsil-Block-Village Panchayat level agencies are engaged in the implementation of National Rural Health Mission Programmes at grassroots level. It has been observed that various rural health programmes (NRHM) are being implemented under the National Rural Health Mission are not properly implemented. The vision, mission, intention, actions of the implementing institutions has been distorted. The real beneficiaries at village level for whom such schemes have been launched are not the real beneficiaries. It has been realized that misappropriate planning and unplanned programme implementation without people's participation has been one of the major reasons responsible for such unwanted result.

An attempt has been made in this study to analyze the problems of health among deprived rural poor multitudes. The problem and prospects of rural health including various programmes being implemented by the various agencies has been properly propagated and the interested rural poor beneficiaries have been motivated to active the government facilities through National Rural Health Mission Programme. An effort has been made to identify, ignored rural poor beneficiaries to make them benefited under the various programmes being implemented under National Rural Health Mission Programme. An innovation of village information system for social infrastructure is only successful when an appropriate rural health services are being properly provided to the needy interested deprived rural poor beneficiaries.

An effort has been made to ensure proper implementation of National Rural Health Mission Programme (NRHM) time to time Rural Health Mela is being organized with the involvement of Social Workers, Medical doctors with the co-ordination of Chief Medical Officer and Primary Health Centre and other interested civil society institutions to achieve the rural health goals in the interest of deprived rural poor multitudes surviving in the remote rural areas.

It is to be noted that the public health initiatives over the years have contributed significantly to the improvement of health services. It is also a fact that the morbidity and mortality level are still unacceptably high. Such unsatisfactory health indices are, in turn, an indication of the limited successes of the public health system in meeting the preventive and curative requirements of the common people surviving in the remote rural areas.

2. Objectives

The main objectives study to analyze the implementation of various rural health programmes being implemented in the village Sarai Khema of Amethi Block.

3. Delivery of Rural Health Programmes

It is evident that in a wide variety of socio-economic settings, rural health programmes have to be designed with enough flexibility, so that such facilities can be delivered to the resourceless down-trodden people living in the remote rural areas. Implementation of the rural health programmes can only be carried out through the appropriate decentralized public health mechanism. Thus broad based rural health initiatives will inevitably continue in the interest of the down-trodden people. It has been observed that the technical and managerial expertise for designing large-span public health programmes exists with the Central Government in a considerable extent. Such expertise can be successfully utilized in designing rural health programmes for implementation in varying socio-economic settings.

Previously the Government has relied upon a 'vertical' implementational structure for the major disease control programmes. However, such an organizational structures has been found extremely expensive and difficult to sustain. In the long time range 'vertical' structure may only be affordable for such disease with other a reasonable possibility for eradication in a foreseeable time.

It has been common perception that over the few years rural health has become vertical structure exclusively for the implementation of family welfare programmes, as a result there is no separate vertical structure and there is no identifiable service delivery system at the grassroots level.

4. Extension of Rural Health Programmes

It has been observed that there is a general shortage of medical personnel in the remote rural areas due to which it has disproportionately impact on the less developed villages. Due to no incentive system, most of the qualified Doctors have induced private medical practices in the rural areas. In the rural health programmes, the effort to deploy qualified medical personnel in such under-served areas has usually been a losing battle under such circumstances the possibility needs to be examined of intrusting some limited rural health programmes functions to nurses, paramedics and other medical personnel from the extended health sector by imparting adequate training to the medical and health workers. There is a vast reservoir of medical practitioners in the rural areas for Medicine, Homoeopathy, who have undergone formal training in their own disciplines. The possibility of using such medical practitioners in the implementation of rural health programmes in order to reach the basic health care in the remote rural areas. It has been observed that State Government has misutilized the financial support under the National Rural Health Mission (NRHM) programme in spite of adopting a policy of devolving programmes and funds in the health sector through different levels of the Panchayati Raj Institution including village Panchayats. The adoption of such an organizational structure has enabled need based allocation of resources and closer supervision through the people's representatives. It needs for a wider adoption of such mode of delivery of rural health programmes in the remote rural areas.

5. Training for Health Care Professionals

It has been observed that the deployment of medical practitioners, staff nurses, in the rural health programmes, is *ad hoc* and significantly short of the requirement for minimal stands of patients care and cure. Medical and Dental education institutions are not evenly spread across various parts of rural areas. Apart from the uneven geographical distribution of the medical institutions, the quality of medical education is highly uneven and in several instances even sub-standard medical quality education has been observed. There has been a common perception that the syllabus is excessively theoretical, making it difficult for the fresh graduate to effectively meet even the primary health care common needs of the down-trodden section of the society. There is general reluctance on the part of qualified medical professionals to serve in the remote rural area as distance from their native/urban place. Certain medical disciplines, such as molecular biology and gene-manipulation, have become relevant in the period after the formulations of the previous policies. The component of medical research have changed radically. In the last seeable future such research will rely increasingly on the new disciplines. It has also been recorded that the current under graduate medical syllabus does not cover such emerging aspects. Certain specialty disciplines, viz. Anesthesiology, Radiology and Forensic Medicine are currently very scarce, resulting in critical deficiencies in the package available in the rural health programmes.

The requirement of expertise in the areas of rural health programmes including 'Public Health' and 'Family Medicine' is markedly more than the expertise required for other clinical specialties. It has been observed that rural health expertise is non-existent in the private health sector, and for short of requirement in the public health sector to bridge this gap Private Public Partnership (PPP) programmes has to be promoted for the training and extension of rural health programmes. It has also been observed that the current curriculum in the graduate/post-graduate course is outdated and unrelated to the contemporary community needs. In respect of Family Medicine, it has to be noted that more talented medical graduates generally seek specialization in clinical disciplines, while the remaining go into genera; practice while the availability of post-gradaute educational facilities is 50 per cent to the total number of qualifying graduates each year and it can be considered adequate, the distribution of disciplines in the post gradaute training facilities is overwhelmingly in favour of clinical specializations. Thus it requires to examine the possible means of ensuring adequate availability of medical personnel with specialization in the 'Rural Health

Programmes' and family medicines to discharge the rural health responsibilities in the remote rural areas.

It has also been observed that the ratio of nursing personnel in the rural areas *vis-à-vis* doctors/beds is very low according to professionally accepted norms. It has also been observed that there is an acute shortage of nurses trained in the super-specialty disciplines for deployment in the rural health programmes in the remote rural areas. A substantial component of rural health programme/primary health care consists of initiatives for disseminating to the citizen, public health related information. The information, education and communication initiatives are adopted not only for disseminating curative guidelines but also as a part of the effort to bring about a behavioural change to prevent the other life style diseases. Rural health programmes, particularly need high visibility at the decentalized level in order to have an impact among the people surviving in the remote rural areas. In reality this task is difficult as 35 per cent of our country's population is illiterate. The existing information, education, and communication strategy is to be fragmented relies to heavily on the mass media and does not address the needs of this segment of the rural population. It has been realized that the effectiveness of Information, Education and Communication programmes are difficult to judge and consequently it is often asserted that accountability for the production use of such financial support is doubtful. Thus the information, education and communication programmes formulated for improving awareness and bringing about behavioural change in the rural people. It has been widely felt and accepted that the school and college students are the most impressionable targets for imparting information related to the basic principles of preventive health care system.

In view of the above Master Social Work (MSW) students of RIWARD Post Graduate College, Hanumanganj, Sultanpur, U.P. attempted to target resourceless deprived rural poor multitude through organizing the "Rural Health Fair" not only to improve the general level of health awareness in regard to "health-promoting" behaviour and MSW students also provide medical support system to the needy and intrusted patients.

6. Environmental and Occupational Health

It has been noted that ambient environmental conditions are significant determinant of the health risks to which as community is exposed.

Unsafe drinking water, unhygienic sanitation and air pollution significantly contribute the burden of diseases. The initiatives in respect of these environmental factors are conventionally undertaken. In this backdrop the efficient implementation of the linked programmes in the health programmes would succeed only to the extent that they are complemented by appropriate programmes in other environment related schemes in the remote rural areas. There is a sub-standard employment due to which workers engaged in such employment become particularly vulnerable to the occupational linked ailments. The long-term risk of chronic morbidity is particularly marked in the case of child labour in the remote rural areas.

7. National Rural Health Mission (NRHM)

Various programmes are implemented under the National Rural Health Mission (NRHM) programmes. Maternity and delivery programme including Janani Suraksha Yojana and Mother Care Programmes are implemented under the National Rural Health Mission Programmes in the Sultanpur. Effort has also been made to upgrade the Primary Health Centre and *ad hoc* appointment of qualified doctors, nursing staff etc. Family Welfare Councillors have also been appointed including the basic health workers at grassroots level to facilitate the need resourceless poor people.

Mother and Child Safety Scheme has also been launched in the District Sultanpur. Mother health programme, safe delivery and mother care programmes have been successfully implemented. The Accredited Social Health Activists (ASHA) and Assistant Nursing Midwife (ANM) have been made responsible for providing monthly report for the pregnant woman and for the pregnant woman registration, Rs. 50 as incentive as are being paid to the ASHA and ANM workers in the rural areas. Village Health Nutrition Day has also been organized to promote the Nutrition and Health Care Programmes. Janani Suraksha Yojana has been launched to control and reduce the death rate. Incentive of Rs. 600 has also been provided to the ASHA workers under the Janani Suraksha Yojana. Saubhagyawati Yojana has also been launched through promoting Private-Public Partnership (PPP) under the National Rural Health Mission Programmes. Immunization and Polio-eradication and other health programmes are being implemented under the National Rural Health Mission Programmes.

In view of the above the Master of Social Work (MSW) students of the RIWARD Post Graduate College, Hanumanganj, Sultanpur have tried to study the Natural-Social-Economic-Institutional Infrastructure of the village and also analyzed ongoing rural development programmes being implemented in the respective village and need based planning of the respective village has also been analysed. MSW students also related 15 Below Poverty Line (BPL) people and provided them the Medical and Health Facilities and Rural Health Fair has also been organized to provide health care and medical treatment to the interested needy patients belonging to the Below Poverty Line and down-trodden section of the society.

8. Conclusion

The rural health is acute problem of the rural poor. Most of the village people are facing critical health problem, which is evident from the fact that in a wide variety of socio-economic settings, rural health programmes have to be designed with enough flexibility, so that such rural health facilities can be delivered to the resourceless deprived rural poor extension of rural health programmes and availability of qualified doctors and nursing staff has been the critical problem of the rural areas for which interest, devotion and dedication of the social workers and health councillors may play an important role in the medical and health services. Educational and training programmes for health care professionals have to be promoted in the remote rural areas.

The requirement of expertise in the areas of rural health programmes including Public Health and 'Family Medicines' is marked by more than the expertise required for other clinical specialties.

Environmental and occupational health has been the acute problem of the villages, unsafe drinking water by hygienic sanitation and air pollution has significantly contributed the burden of diseases. National Rural Health Mission (NRHM) Programmes have been launched in the rural areas. Janani Suraksha Yojana, Mother and Child Care, Saubhagyawati Yojana, immunization and polio-eradication and various other rural health programmes have been launched to facilitate needy patients.

Master of Social Work (MSW) Students of RIWARD Post Graduate College, Hanumanganj, Sultanpur (U.P.), India have been engaged in providing requisite health support to the interested needy patients under the National Rural Health Mission (NRHM) programme which has been appreciated at wider extent by the village people. Thus it can be said that social workers and Rural Health Councillor may play an important role in the successful implementation of rural health programmes for the resourceless deprived down-trodden section of the society.

22

Vehicular Pollution: Health Effect and Its Control

Manish Yadav and *Uttara Yadav*

Introduction

Air pollution is the presence of undesirable material in air, in quantities large enough to produce harmful effects. This definition does not restrict air pollution to human causes, although we normally only talk about these. The undesirable material may damage human health, vegetation, human property, or the global environment as well as create aesthetic insults in the form of brown or hazy air or unpleasant smells (Nevers, 1993).

Most of the Indian cities are also experiencing rapid urbanization and the majority of the country's population is expected to be living in cities within a span of next two decades. Since poor ambient air quality is largely an urban problem this will directly affect millions of the dwellers in the cities.

The rapid urbanization in India has also resulted in a tremendous increase the number of motor vehicles. The vehicle fleets have even doubled in some cities in the last one decade. This increased mobility, however, comes with a high price. As the number of vehicles continues to grow and the consequent congestion increases, vehicles are now becoming the main source of air pollution in urban India.

In Delhi, the data show that of the total 3,000 metric tonnes of pollutants emerged out everyday, close to two-third (66%) is from vehicles. Similarly, the contribution of vehicles to urban air pollution is 52 per cent in Bombay and close to one-third in Kolkata. It has been estimated that in Santiago, Chile, wherever pollution concentration exceeds ambient standards, mobile sources or vehicles are the cause. Similarly, in case of Budapest, Hungary, transport is the dominant source of emissions except sulphur dioxide (SO_2), contributing 57 per cent of Oxides of Nitrogen (NO_x), 80 per cent of lead (Pb), 81 per cent of carbon monoxide (CO) and 75 per cent of hydrocarbon (HC) emissions (Lehoczki, 2000).

Although, the air quality can be improved through a combination of technical and non-technical measures, legislative reforms, institutional approaches and market-based instruments, there are certain unique challenges which the country has to face in tackling the problem of urban air pollution.

The document presents a review of the vehicular emission problems in Indian cities, the various developments that have taken place in the past including the studies conducted for assessment of the air quality in cities, the legislation and standards adopted for the control of vehicle emissions, the role of the various concerned agencies, the steps taken for improvement in the quality of the automotive fuel, the overall impact of these measures and the future strategy to be adopted for vehicular emission reduction and related issues.

Major Vehicular Pollutants

One of the major issues associated with road transport today is poor air quality. The drastic increase in number of vehicles has also resulted in a significant increase in the emission load of various pollutants. Of recent interest is exposure to high atmospheric levels of particulate matter (PM), sulphur oxide (SO_x), hydrocarbons, nitrogen oxide (NO_x), carbon mono oxide (CO), carbon di oxide (CO_2), ozone (O_3), and many heavy metals which have been found to be associated with negative health impacts. Gasoline and diesel vehicles and are found to be a significant source of these pollutants.

Pollutants in Gasoline Vehicles' Emissions

Emissions from Gasoline Vehicles primarily consisting of CO, HC, oxides of nitrogen (NO_x), and partial oxides of aldehydes, besides particulate matters including Pb salts account for the larger chunk of all pollution from gasoline-run vehicles. Though now-a-day's lead is replaced by benzene so there is no more concern remains due to lead.

Pollutants in Diesel Vehicles' Emissions

As diesel engines breathe only air, blow by gases from the crankcase (consisting primarily of air and HC) are rather low. Moreover, due to its low volatility, evaporative emissions from the fuel tank can safely be ignored. Though the concentration of CO and un-burnt HC in the diesel exhaust are rather low, they are compensated by high concentration of NO_x and SO_x (higher than that in gasoline vehicles). There are smoke particles and oxygenated HC, including aldehydes and odour-producing compounds which have high nuisance value.

Effect of Vehicular Pollution

The vehicular emissions have damaging effects on human health, ecology, monuments and environment. There is a wide range of adverse health/environmental effects of the pollutants released from vehicles. The effects may be direct as well as indirect covering right from reduced visibility to cancer and death in some cases of acute exposure of pollutants specially carbon monoxide. These pollutants are believed to directly affect the respiratory and cardiovascular systems. In particular, high levels of sulphur dioxide and Suspended Particulate Matter are associated with increased mortality, morbidity and impaired pulmonary function. Air quality and climate quickly changed in the last years, requiring new researches focused on pollution effects, not only on environment and human health but also on outdoor Cultural Heritage. Vehicular pollution also contributing in emission of greenhouse gases in this section we are going to discuss in detail the effect of vehicular pollution on health, plants, monuments and in generation of greenhouse emission.

Effect on Human Health

Heavy transportation in metropolitan cities is a major contributor to environmental pollution in addition to industrial and commercial activities. The population residing in the vicinity, daily commuters, and business people are always exposed to the traffic air containing particulate matter (<10 lm; PM10), inorganic gases, and volatile and semi-volatile organic compounds. Several studies reported significant health risks and increased morbidity and mortality rates and hospital admissions because of cardio respiratory diseases, oxidative stress, and an increase in the incidence of cancer among the urban population. The submicronic airborne particulates have an aerodynamic diameter of less than 2.5 lm. The particulates result from incomplete combustion of diesel and petroleum fuels. As per the study results of the American Cancer Society (ACS), an average increase of 10 lg/m^3 of PM10 or PM2.5 was associated with a 3 per cent-4 per cent increase in mortality.

Vehicular pollution is often associated with respiratory and cardiovascular diseases in children, but few studies have explored its neurodegenerative effects in a developing country like India. Attention-deficit/hyperactivity disorder (ADHD) is one of the most common behavioural changes diagnosed in children and adolescents. It is characterized by inattentiveness, overactivity, and impulsiveness (Nuhoglu, 2005; Woolhouse, 1986; Bansal, 1988). Children with ADHD are forgetful, restless, prone to fail, unable to follow tasks, unpredictable, and moody. The problem starts in early childhood and persists in adulthood in the majority of cases. The disorder is characterized by marked inattention, hyperactivity, and impulsiveness. In most cases, symptoms can be treated by catecholamine-releasing drugs. Children with ADHD are at a greater risk of substance abuse, conduct, and mood disorders.

Effect on Plants

In recent past, air pollutants, responsible for vegetation injury and crop yield losses, are causing increased concern SO_2 (Fuji, 1973), which is a major component of air pollutants, affects morphological characteristics of plants such as number of leaves, leaf area, length of stem and roots and number of flowers and fruits (Wali, 2004). Air pollutants also affect germination of seeds, length of pedicles, and, number of flowers in inflorescence (Nithamathi, 2005). Several studies have shown the impact of automobile exhaust on road side vegetation through their visible and non-visible effects (Farmer and Lyon 1977; Fleckiger *et al.*, 1978; Keller, 1974; Sarkar *et al.*, 1986; Leon, 1988; Nuhoglu, 2005). One of the most common impacts is the gradual disappearance of chlorophyll and concomitant yellowing of leaves, which may be associated with a consequent decrease in the capacity for photosynthesis (Woolhouse, 1986).

The photosynthetic pigments are the most liable to be damaged by auto pollution. It was noticed that chlorophyll *a* content decreased with increasing pollution load. A wide variation in chlorophyll *b* content was also observed in the foliar tissue of plants growing at different sites in relation to the intensity of urban pollution due to automobiles (Amitosh Verma, 2006).

Chlorophyll is said to be an index of productivity, hence any alteration in chlorophyll concentration may change the morphological, physiological and biochemical behaviour of the plant. Air pollution-induced degradation in photosynthetic pigments was also observed by a number of workers (Bansal, 1988; Singh *et al.*, 1990).

Overview of the Emission Norms in India

1991: Idle Co. Limits for Gasoline Vehicles and Free Acceleration Smoke for Diesel Vehicles, Mass Emission Norms for Gasoline Vehicles.

1992: Mass Emission Norms for Diesel Vehicles.

1996: Revision of Mass Emission Norms for Gasoline and Diesel Vehicles, Mandatory Fitment of Catalytic Converter for Cars in Metros on Unleaded Gasoline.

1998: Cold Start Norms Introduced.

2000: India 2000 (Eq. to Euro I) Norms, Modified IDC (Indian Driving Cycle), Bharat Stage II Norms for Delhi.

2001: Bharat Stage II (Eq. to Euro II) Norms for All Metros, Emission Norms for CNG and LPG Vehicles.

2003: Bharat Stage II (Eq. to Euro II) Norms for 11 Major Cities.

2005: From 1st April Bharat Stage III (Eq. to Euro III) Norms for 11 Major Cities.

2010: Bharat Stage III Emission Norms for 4-Wheelers for Entire Country whereas Bharat Stage-IV (Eq. to Euro IV) for 11 Major Cities.

Overview of the Fuel Quality in India

In India we are yet to address the vehicle and fuel system as a whole. It was in 1996 that the Ministry of Environment and Forests formally notified fuel specifications. Maximum limits for critical ingredients like Benzene level in petrol have been specified only recently and a limit of 5 per cent m/m and 3 per cent m/m has been set for petrol in the country and metros respectively. In place of phase-wise upgradation of fuel specifications there appears to be a region-wise introduction of fuels of particular specifications. The high levels of pollution have necessitated eliminating leaded petrol, throughout the country.

To address the high pollution in 4 metro cities 0.05 per cent sulphur petrol and diesel has been introduced since 2000-01. The benzene content has been further reduced to 1 per cent in Delhi and Mumbai. While unleaded gasoline introduced in 1995 and 100 per cent in 2000 earlier (Emission Controls Manufacturers Association, 2011).

Control of Emission through Vehicular Pollution

For containing vehicular pollution, the Government has taken important initiatives in recent years. The Union Government and the Provincial Governments in India have been emphasizing the need for planning and developing strategies to implement mitigation measures to maintain the urban air quality and make the cities cleaner and greener for achieving better air quality and good health for its citizens. Over the past decade or so, the government has bought in statutes aimed at regulating and monitoring industrial and vehicular pollution across the country.

Technical and Non-Technical Measures

Vehicular Pollution Control Initiatives in India can broadly be categorized into Technical and Non-Technical Measures. Technical Instruments for controlling vehicular pollution includes implementation of stringent emission norms for both new and In-Use Vehicles, Improvement in Vehicular Technology, Improvement in the Quality of fuels, Switching over to cleaner vehicles as well as fuels, Implementation of I/M system etc. While Non Technical instruments like Better Traffic Management system, Augmentation in public transport system, Implementation of Market based instruments, i.e. fiscal instruments, generating mass awareness, drives for checking adulteration etc. Forthcoming section will discuss in details about various measures those have been taken all over the country for controlling this menace of Vehicular Pollution.

Introduction to the Technology for Emissions Control

Catalytic Converters

The term covers the stainless steel box mounted in the exhaust system. Inside is the autocatalyst—a ceramic or metallic substrate with an active coating incorporating chemical compounds (the wash coat) to support a combination of catalytical materials or minerals selected for their effectiveness in the required emissions reductions. It can also be an homogeneous honeycomb ceramics in which only active compounds are extruded simultaneously. The autocatalyst is mounted in a can and is protected from vibration and shock by a resilient 'mat'. The catalytic converter then looks similar to an exhaust muffler.

Particulate Filters

Based on engine technology and application specificities, different filter technologies may be used to reduce particles emissions. The 'wall-flow' filter is a ceramic substrate in which the gas is forced to flow through the walls, thus filtering out particulate matter with more than 99 per cent efficiency. It is also mounted in a stainless steel can and protected by a resilient mat. These filters can also have an active coating similar to that used in autocatalysts to ensure their regeneration. Fuel-borne catalysts or engine calibration measures can also be used for the necessary regeneration.

The 'partial-flow' filter is a device which typically separates 30 to 60 per cent of the particulates from the exhaust gas. These filters are available in various materials, from metallic to fibre-based.

Particulate filters are generally used with diesel engines to remove diesel particulate matter (PM), but in principle can be used with other types of engine/fuel combinations, although these produce less particulate matter. Based on engine technology and application specificities different filter technologies may be used to reduce particles emissions (Emission Controls Manufacturers Association, 2011).

Traps and Adsorbers

Traps and adsorbers are related to autocatalysts but are used to control the emissions of specific pollutants—usually NO_x or HC—where some operating conditions or exhaust composition are unsuitable for continuous catalytic action. They store the pollutant for a period of time but then release it when conditions are right for it to react over the catalytic materials, reducing the emissions (Emission Controls Manufacturers Association, 2011).

Substrates

The substrates on which the active catalyst is supported can be ceramic or metallic, with each offering particular advantages for specific applications and positions in the exhaust system. The development of strong ultra-thin wall substrates with cell densities of up to 1200 cells per square inch (186 cells/square cm) has been a major factor in the increasing efficiency of autocatalysts (Emission Controls Manufacturers Association, 2011).

Catalytic Coatings

Coatings systems have been developed which allow the maximum efficiency with optimum use of the precious metals platinum (Pt), palladium (Pd) and rhodium (Rh). The nanotechnology used in catalytic coatings involves stabilised crystallites, washcoat materials that maintain high surface area at temperatures around 1000°C, improved oxygen storage components and novel coating processes to optimise the distribution of the coatings. All play a part in the high efficiencies of autocatalysts (Emission Controls Manufacturers Association, 2011).

Emission Standards for In-Use Vehicles and I/M Practice

Inspection and Maintenance (I/M) Practice in India

In major cities there exists a mandatory system for inspection and maintenance but it is now increasingly felt to upgrade the present system to a more effective one in the near future. Various studies conducted worldwide have indicated that with the implementation of proper I/M system there could be 30-40 per cent reduction in pollution load from vehicles. Every commercial vehicle in India has to go for a mandatory fitness test. The renewal period for fitness certification in general is 2 years for new commercial vehicle and every 1 year for old vehicles. For private vehicles no mandatory periodic fitness check is required in India but there exist a system of re-registration of private vehicles after 15 years of initial registration. But all in-use vehicles are required to obtain emission check certificate called Pollution Under Control (PUC). Frequency of this PUC certification varies from 2 to 4 times a year. This PUC is issued based on conformity to idle emission test for gasoline vehicles and free acceleration smoke test for diesel vehicles, (CPCB, Status of the Vehicular Pollution Control Programme in India, 2010).

Pollution under Control (PUC) Certification System

The first initiative to regulate vehicle emissions in India started in the year 1989 when Ministry of Environment and Forests constituted an expert committee to notify the emission standards for both new and in-use vehicles under the Environment (Protection) Act. The in-use vehicle emission norms have been tightened with effect from 1st October, 2004 and computerization model for emission check has been developed by Society of Indian Automobile Manufactures (SIAM), which is already in place in the major metro cities.

Improvement in Fuel Quality

Much of the pollution control depends on the quality of the fuel. There are various constituents/parameters in the fuels (Gasoline and Diesel) together contributes towards emissions from the vehicles. There exist code of standards or specifications for the gasoline and diesel to be sold in the market in every country. National or other legally enforceable specifications represent the minimum quality that must be supplied and it is implicit that engine designers should ensure that their vehicles would run satisfactorily on such a quality of fuel. In India the Bureau of Indian Standards (BIS) notifies the requisite specifications for gasoline and diesel. There are a total of 15 parameters in gasoline specifications out of which 4 are environment related parameters. Similarly, there is a total 16 parameters in diesel specifications, out of which 4 are environment related parameters.

Use of Alternate Fuel

The selection of alternate fuels depends on the load of emission to be allowed, the technology available and the cost of system to be developed. A system that fulfils the legislative requirements and can be sold at the lowest price is to be generally accepted. The Alternate fuels have to meet the following criteria:

1. Technical Acceptability
2. Economically Competitiveness
3. Environmentally acceptability
4. Safety and Availability

The important fuels that are considered as meeting the above criteria include Natural Gas (CNG/LNG), Propane (LPG), Ethanol, Methanol, Diesel, Electric fuel, Hydrogen, Di-methyl Ether (DME), P-series Fuel Cell and Solar fuels. Indian Government has taken various initiatives time to time for the development and promotion of cleaner alternative to conventional automotive fuels, i.e. diesel and gasoline.

Conclusion

The ambient air quality keep on deteriorating with rapid urbanization and major contributor for that is increase in the number of vehicle. These vehicles emits some harmful gases which have got serious ill effects on human health and plants.

Though all these emission can be controlled if use the technology discussed above and it is seen also that air of Delhi has been changed a lot in past recent years. The only thing is responsible for that is Judicial system and the Government who implement the CNG strictly. So the role of Government, Judiciary, manufacturs and at last role of public is important for the control of pollution causing due to vehicles.

REFERENCES

Agarwal, A. (1999). *Engine of the Devil.* New Delhi: Centre for Science and Environment.

Amitosh Verma, S.S. (2006). Biochemical and Ultrastructural Changes in Plant Foilage Expoosed to Auto Pollution. *Environmental Monitoring and Assessment* 120, 585-602.

Ankur, N.A. (2010, April 1-15). Metrocity. *Down to Earth.*

B.P. Chattopadhyay, A.M. (2007). Exposure to Vehicular Pollution and

Assessment of Respiratory Function in Urban Inhabitants. *Lung,* 185, 263-270.

Bansal, S. (1988). Studies on the effect of certain atmospheric pollutants on fruit diseases of Lycopersiconesculentum Mill. Caused by Alternaria alternata. Ph.D. Thesis. Bhopal Univ. Bhopal.

Bell, U.N. (2005). Who Changed Delhi's Air? The Roles of the Court and the Executive in Environmental Policymaking. *Resource for the future,* 05-48.

Chattopadhyay, V. (2010, Feb. 1-15). Electric Buzz. *Down to Earth.*

CPCB. (2004-2005). *Automobile Pollution, Annual Report Ch.5.*

CPCB. (2010). *Status of the Vehicular Pollution Control Programme in India.* New Delhi.

F. Dursbeck, L. Erlandsson, C. Weaver. (2001). Status of Implementation of CNG as a Fuel for Urban Buses in Delhi. *Centre for Science and Environment.*

Fuji, S. (1973). The current damage of plant damage by air pollution in the Ckayama Prefecture. *Shokbutsu Boeki, 27,* 249-252.

Gas, M. O. (2003). *Auto Fuel Policy.*

Heywood, J.B. (1988). *Internal Combustion Engine Fundamentals.* McGraw Hills.

Jacobson, M. (2005). Effects of biofuels vs. other new vehicle technologies. *Int J. Biotecnology,* Vol. 11.

Kathuria, V. (2005). *Vehicular Pollution Control—Concept note.* Chennai.

Manab Das, S.K. (2006). Distribution of Pm2.5 And Pm10-2.5 In Pm10 Fraction in Ambient Air due to Vehicular Pollution in Kolkata Megacity. *Environmental Monitoring and Assessment 122,* 111-123.

Nevers, N.D. (1993). *Air Pollution Control Engineering.* McGraw Hill.

Nithamathi, C.P. (2005). Impact of air pollution on Ceasalpinia sepiaria Linn. in Tuticorin city. *Indian Journal of Environment and Ecoplanning,* 10(2), 449-452.

Nuhoglu, Y. (2005). The harmful effects of air pollutants around the Yenikoy thermal power plant on architecture of Calabarian pine (Pinus brutila Ten.) needles. *Journal of Environmental Biology,* 26, 315-322.

P.C. Joshi, A.S. (2007). Physiological Responses of Some Tree Species under Roadside Automobile Pollution Stress Around City of Haridwar, India. *Environmentalist,* 365-374.

Padhy, P.K., V.C. (2000). Total non-methane volatile organic compounds (TNMVOC) in the atmosphere of Delhi. *Atmospheric Environment,* 577-584.

Rao, N.M., P.T. (1991). A dose-response relationship between pollution index and pulmonary function in shopkeepers exposed to auto exhaust. *Indian J. Environ Protect* 11(10), 737-740.

Region, E.P. (2001). *Report on clean fuels in responce to Hon'ble Supreme Court Order.*

S.P. Chavan, A.M. (2010). Effect of vehicular pollution on plants: A case study from Aurangabad city (MS), India. *2nd International Conference on Chemical, Biological and Environmental Engineering.*

S. Siddique, M.B. (2011). Attention-Deficit Hyperactivity Disorder in Children Chronically Exposed to High Level of Vehicular Pollution. *Eur J. Pediatr 170*, 923-929.

S.K. Goyal, S.G. (2006). Understanding Urban Vehicular Pollution Problem *vis-à-vis* Ambient Air Quality—Case Study of a Megacity (Delhi, India). *Environmental Monitoring and Assessment* 119, 557-569.

Srivastava, A., J.A. (2005). Air toxics in ambient air of Delhi. *Atmospheric Environment* 39, 59-71.

Standards, B.O. (2011). *Programme of work.*

Wali, B.M. (2004). Plant growth, stomatal response, pigments and photosynthesis of Althea officinalis as affected by SO_2 stress. *Indian Journal of Plant Physiology*, 224-233.

Woolhouse, H.Y. (1986). *Procesa and control of plant senescence.* Elsevier Science Publisher.

23

Role of National Rural Health Mission (NRHM) in Providing Effective Health Care Services to Rural Population in India

Rajneesh Chaudhary

Introduction

Recognizing the importance of health in the process of economic and social development and improving the quality of life of our citizens, the Government of India has launched the National Rural Health Mission to carry out necessary architectural correction in basic health care delivery system, with the objective to reduce infant and maternal mortality, prevention of communicable and non-communicable diseases, population stabilization, and universal access to public health services such as water sanitation, hygiene, immunization, nutrition and promotion of healthy life styles.

NRHM seeks to provide effective care to rural population throughout the country especially the disadvantaged groups by improving access, enabling community ownership and demand for services, strengthening public health system for efficient service delivery system. It covers the entire country with especial focus on 18 states of India. It is designed to galvanize the various components of primary health system like preventive, promotive and curative care, human resource management, diagnostic services, logistic management, disease management, and surveillance for improved service delivery.

In the last four years this mission has achieved its goal to a great extent. As per its predetermined goals, i.e. reduction of IMR to 30/

1000 live births, reduction of maternal mortality ratio to 100/100000 live births and reduction of TFR to 2.1 are to be achieved by 2010. Against the above goals, the IMR of 55/1000 live births and MMR of 254/100000 live births and TFR of 2.7 have been achieved. *Janani Suraksha Yojana* and ASHA are working actively in the field to connect household with health facilities, village health and sanitation committees are constituted. This programme also increased the human resource in the health system. Programme is also giving emphasis upon revitalizing the Indian system of medicine, community monitoring initiatives and partnership with NGOs, school health programmes etc.

Objective

The objective of this paper is to bring forward the role of National Rural Health Mission in providing effective health care facilities and of those who are playing an important role in improving the health status of rural population in India.

Introduction

India is the second most populous country of the world and has changing socio-political demographic and morbidity patterns that have been drawing global attention in recent years. Despite several growth orientated policies adopted by the government, the widening economic, regional and gender disparities are posing challenges for the health sector. About 75 per cent of health infrastructure, medical man power and other health resources are concentrated in urban areas where 27 per cent of the population live and rest of the population reside in rural areas. Contagious, infectious and waterborne diseases such as diarrhoea, amoebiasis, typhoid, infectious hepatitis, worm infestations, measles, malaria, tuberculosis, whooping cough, respiratory infections, pneumonia and reproductive tract infections dominate the morbidity pattern, especially in rural areas. However, non-communicable diseases such as cancer, blindness, mental illness, hypertension, diabetes, are also on the rise. The health status of Indians is still a cause for grave concern, especially that of the rural population. This is reflected in the life expectancy (63 years), infant mortality rate (63/1000 live births), maternal mortality rate (417/100000 live births); however, over a period of time some progress has been made.

To improve the prevailing situation, the problem of rural health is to be addressed both at macro (national and state) and micro (district and regional) levels. This is to be done in an holistic way, with a genuine effort to bring the poorest of the population to the centre of the fiscal policies. A paradigm shift from the current 'biomedical model' to a 'socio-cultural model', which should bridge the gaps and improve quality of rural life, is the current need. A revised National Health Policy addressing the prevailing inequalities, and working towards promoting a long-term perspective plan, mainly for rural health, is imperative.

The more recent experiences of a few developing countries (Sri Lanka) and state of Kerala in India illustrate the way in which health forms part of development, this was because the efforts made in the field of health were simultaneously reinforced by development in the other sector such as education, social welfare and land reforms. The link between health and development has been clearly established, one being the starting point for the other and *vice versa*. Since health is integral part of development.

Recognizing the importance of health in the process of economic and social development and improving the quality of life of our citizens, the Government of India has launched the National Rural Health Mission in 2005 to carry out necessary architectural correction in basic health care delivery system. The Government of India made a commitment to raise the public expenditure on health from 1 per cent of GDP to 2-3 per cent of GDP over the mission period (2005-2012).

NRHM seeks to provide effective care to rural population throughout the country especially the disadvantaged groups including women and children by improving access, enabling community ownership and demand for services, strengthening public health system for efficient service delivery, enhancing equality and accountability and promoting decentralization. NRHM covers the entire country with especial focus on 18 states—Arunachal Pradesh, Uttar Pradesh, Assam, Bihar, Himachal Pradesh, Jharkhand, Jammu and Kashmir, Manipur, Mizoram, Madhya Pradesh, Tripura, Sikkim, Orissa, Rajasthan, Uttarakhand, and Meghalaya.

The National Rural Health Mission aims to provide for an accessible, affordable acceptable and accountable health care through a functional public health system. The goal of mission is to improve the availability of and access to quality health care by people especially for those residing in rural areas the poor, women and children. It is designed to galvanize the various components of primary health system like preventive, promotive and curative care, human resource management, diagnostic services, logistic management, disease management, and surveillance, and data management systems etc. for improved service delivery.

The Vision of this Programme is

- To increase public spending on health from 0.9 per cent GDP to 2-3 per cent GDP with improved arrangement for community financing and risk pooling.
- To undertake architecture of the health system.
- Effective integration of health concerns through decentralized planning at districts, with determinants of health.
- To revitalize local health traditions and mainstream AYUSH into the public health system.
- To address inter-state and inter-district disparities.
- To improve access to rural people especially poor women and children to equitable affordable, accountable and effective primary health care.

The Objectives of the Mission are

- Reduction in IMR and MMR.
- Universal access to public health services such as women's health, child health, water, sanitation and hygiene, immunization and nutrition.
- Prevention and control of communicable and non-communicable diseases, including locally endemic diseases.
- Access to integrated comprehensive primary health care.
- Population stabilization, gender and demographic balance.
- Revitalize local health traditions and mainstream AYUSH.
- Promotion of healthy life styles.

The Main Approaches of NRHM

1. To Communitize

- Decentralized planning.

- Formation of hospital development committees/PRIs at all levels.
- Inter-sectoral convergence.
- Formation of village health committees.

2. To Provide Improved Management through Capacity Building

- To improve management skills of Block and District health officers through capacity building.
- Participation of NGOs in capacity building.
- Continuous capacity building programmes for development of skills.

3. To Provide Flexible Financing

- Making available united grants to health institutions.
- Participation of NGOs for public health goals.
- To provide more resources for more reforms.

4. Innovation in Human Resource Management

- Insuring availability of locally resident health workers.
- 24×7 medical emergency at CHC.
- Integration with AYUSH.
- Selection of ASHAS, ANM, AWW, HW.

5. To Monitor Progress against the Standards

- Formation of independent monitoring committees at Block, District and State level.
- Preparation of annual reports.

Status of Public Health in India: Progress under NRHM

In the last four years this mission has achieved its goal to a great extent. As per its predetermined goals such as:

Development of Infrastructure and Manpower

The programme has strengthened 30,000 PHCs with 3 staff nurse each, 6500 CHCs strengthened with 7 specialists and 9 staff nurse, out of those 7348 PHCs and 2720 CHCs are working for 24 × 7, about 1.5 lakh sub-health centres established with 2 ANMS, 600 district hospitals and 1800 sub-divisional hospitals strengthened to provide quality health services.

Rogi Kalyan Samiti (RKS), village health sanitation committee, state and district health society, Janani Suraksha Yojana, mobile medical units are also making a remarkable impact on community. 23100 Rogi Kalian Samitis established in all district hospitals, PHCs and CHCs, 304 districts have functional Mobile Medical Units, village health sanitation committee constituted in 6 lakh villages and united grants provided to them.

The human resource is also increased such as the appointments of AYUSH doctors besides MBBS doctors. Over 7 lakh trained ASHAs working actively in the field to connect households with health facilities.

IMR Reduced to 30/1000 Live Births by 2012. Material Mortality Reduced to 100/100,000 by 2012. TFR Reduced to 2.1 by 2012

IMR was 58/1000 live births in 2005. It is down to 55/1000 live births (SRS, 2007). There has been a 4-point decline in IMR. More concerted efforts to tackle malnutrition and neonatal mortality will facilitate a 5-point decline required for achievement of expected outcome (see Fig. 23.1).

Reduction of Maternal Mortality to 100/100000 by 2012

MMR was 301 in 2001-03. It was 254 in 2004-06. Janani Suraksha Yojana was launched in 2005 and early gains are captured in the reduction the thrust on institutional deliveries and assured referral transport together with efforts to improve the quality of care in facilities is likely to further increase the pace of reduction of MMR (see Fig. 23.2).

TFR Reduced to 2.1 by 2012

TFR was 2.9 in 2005. It reduced to 2.6 in 2008. 14 States/UTs are already below replacement level. 7 States and UTs are between 2.2 and 2.6, Bihar, UP, MP, Rajasthan, Jharkhand and Chhattisgarh have TFR between 2.7 and 3.9 (see Fig. 23.3).

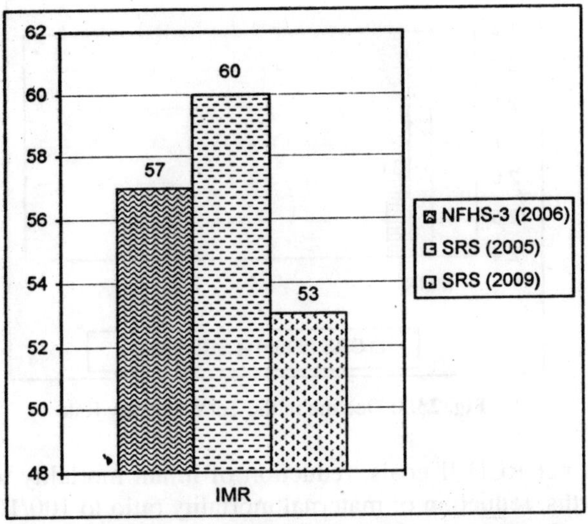

Fig. 23.1: Significant Decrease in Infant Mortality Rate of India

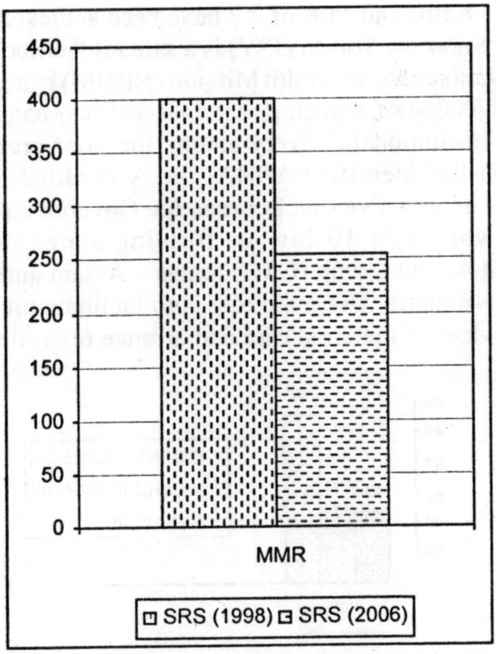

Fig. 23.2: Decrease in the Maternal Mortality Ratio of India

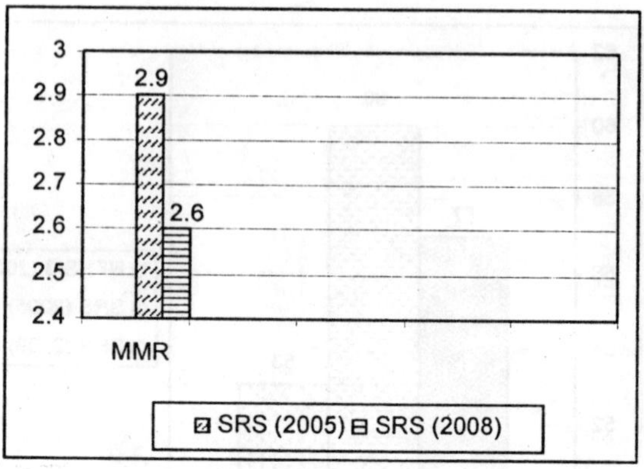

Fig. 23.3: Decline in the Total Fertility Rate

As per RCH-II goals, reduction of infant mortality to 30/1000 live births, reduction of maternal mortality ratio to 100/100000 live births and reduction of TFR to 2.1 are to be achieved by 2010. Against the above goals, the IMR of 55/1000 live births and MMR of 254/ 100000 live births and TFR of 2.7 have been achieved (SRS, 2007).

Janani Suraksha Yojana (JSY) is a safe motherhood intervention under the National Rural Health Mission (NRHM) being implemented with the objective of reducing maternal and neo-natal mortality by promoting institutional delivery among the poor pregnant women. The Yojana has identified ASHA, the Accredited Social Health Activists as an effective link between the Government and the poor pregnant women in 10 low performing states, namely eight Empowered Action Group (EAG) States – Assam and J&K and the remaining NE States. Her main role is to facilitate pregnant women to avail services of maternal care and arrange referral transport.

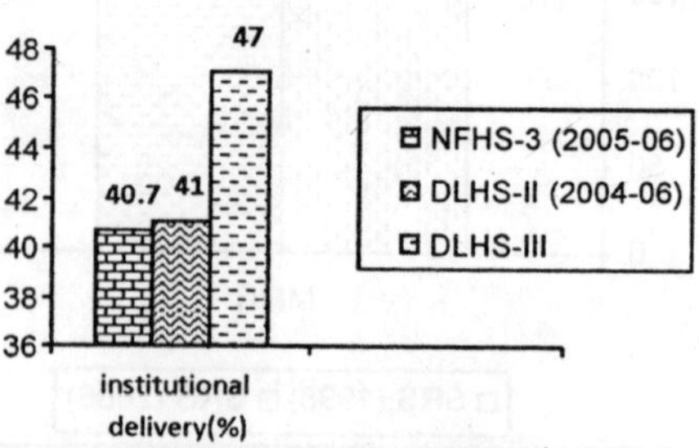

Fig. 23.4: Showing Trends in Institutional Delivery

The scheme focuses on the poor pregnant women with special dispensation for states having low institutional delivery rate, namely the states of Uttar Pradesh, Uttrakhand, Bihar, Jharkhand, Madhya Pradesh, Chhattisgarh, Assam, Orissa, Rajasthan and Jammu and Kashmir. While these states have been classified as Low Performing States (LPS), the remaining states have been named as High Performing States (HPS). Besides the maternal care, the scheme provides cash assistance to all eligible mothers for delivery care. A considerable progress has been made in JSY. 7.04 lakh women covered under JSY in 2005-06, the coverage in 2008-09 was 86.22 lakh and 78.41 lakh in the year 2009-10.

National Vector Borne Disease Control Programme

The National Health Policy has set the goals for reduction of Malaria Mortality Reduction Rate by 50 per cent up to 2010, additional 10 per cent by 2012, Kala Azar Mortality Reduction Rate – 100 per cent by 2010, sustaining elimination until 2012, Filaria/Microfilaria Reduction Rate – 70 per cent by 2010, 80 per cent by 2012 and elimination by 2015 and Dengue Mortality Reduction Rate – 50 per cent by 2010 and sustaining at that level until 2012.

NRHM has made a remarkable reduction in the incidence and prevalence of these vector borne diseases. 45.23 per cent reduction in malaria mortality reported in first two years (2006 to 2008). There is an issue of under reporting of cases which also needs to examine. 21.93 per cent reduction in deaths due to Kalaazar from 2006 to 2008. During the year 2007, 44533 cases and 203 deaths whereas during the year 2008, 32081 cases and 141 deaths were reported. During the year 2009 (up to March, 2009) 2254 cases and 5 deaths have been reported. 56.52 per cent reduction in Dengue mortality in first two years.

Leprosy Prevalence Rate—Reduce less than 1 per 10,000 thereafter: Target achieved in December 2005 and maintained thereafter.

Tuberculosis DOTS series—Maintain 85 per cent cure rate through entire mission period: 87 per cent cure rate has been maintained. Case detection rate has been moved from 70 per cent to 72 per cent. Due to successful implementation of RNTCP, prevalence of all forms of TB has been brought down from 586/lakh population (1990) to 283/lakh population in 2007 and TB mortality in the country has reduced from over 42/lakh population in 1990 to 28/lakh population in 2007 as per the WHO global report 2009. Repeat population surveys indicate an annual decline in prevalence of disease by 12 per cent.

Upgrading Community Health Centres according to Indian Public Health Standards (IPHS): Physical infrastructure upgradation human resource augmentation, equipment provision taken up in nearly all community health centres DLHS-III found 90.1 per cent having normal delivery services since the IPHS provides for a high standard it will take some time before augmentation is as per IPHS, however, service guarantees have shown considerable improvement in the common review missions.

Increase utilization of First Referral units from less than 20 per cent to 75 per cent: Though no separate data on utilization levels in FRU is currently available the common review mission has reported much higher utilization of in-patient facilities due to increased institutional deliveries.

Engaging Accredited Social Health Activists (ASHAs) in 18 states: Fully trained Accredited Social Health Activists (ASHAs) for every 1000 population in 18 special focused states. 7 lakh ASHAs in all states have been selected, 5.81 completed training up to 5th module. ASHAs are working in the field to connect households with health facilities.

Public health is a marathon and 5 years is a very short time. In spite of this, NRHM has made a remarkable impact on the public system of health care in the country.

24

Role of Availability and Beneficiaries of Inclusive Village Development Programme in Reproductive Health Services in Uttar Pradesh, India

Diwakar Yadav and *Shashi Kant Srivastava*

Background

The question is remained unanswered, does availability and beneficiaries of an inclusive (health and rural) development programme enhance the utilization of maternal-child health and contraceptive use? Does synergy between exclusive development programmes have an effect on maternal-child health care and family planning (MCH-FP) service utilization in Uttar Pradesh?, become critical to execute the programme in efficient manner. The answer to above questions holds important for advocating "healthy people, healthy village, and healthy nation".

More recently developed demographic perspectives have implicitly assumed that the consequences of population growth are serious and problematic, and they more directly to explanations of the causes of population growth. The demographic transition theory suggested that growth is an intermediate stage between the more stable conditions of high birth and death rates to a new balance of low birth and death rates. Reformulations of the demographic transition perspectives have emphasized its revolutionary character and have shown that the demographic transition is not one, monolithic change, but rather that it encompasses several interrelated transitions: A decline in fertility, and by subsequent transitions in migration, urbanization, the age structure, and the family and household structure in society.

India is trying to reach in fourth stage of demographic transition in terms of low birth and death rates. India was the first country in the world to launch the family planning programme (FPP) in 1951 with a focus on population growth and subsequently launched area specific rural development programme. This FPP was later renamed family welfare programme in 1977 integrating the maternal child health component, realizing the positive association between low birth rate and improved infant child survival. India further modifies the maternal and child health (MCH) programme to child survival and safe motherhood programme (CSSM) in 1992. This programme was later integrated into the reproductive and child health (RCH) programme in 1998 to improve maternal child health and to meet the needs of family planning services. After RCH programme, Government of India launched a flagship programme National Rural Health Mission (NRHM) 2005-2012, focuses on providing effective health care to the rural population throughout the country, with special emphasis on eighteen states with weak public health infrastructure. The NRHM attempts to synergise health issues with determinants of health, like sanitation and hygiene, nutrition, and safe drinking water. It also aims at improving the availability, accessibility, affordability and quality of effective health care services to rural population particularly underserved women and children.

The integration of family planning programmes and maternal child health care services are not new. In the early 1980s, various empirical evidences from experimental studies of developing countries were introducing the integration of family planning programmes with the subsequent development programmes (Faruqee, 1982; Rahman, *et al.,* 1980; Reinke, 1985; Taylor *et al.,* 1983). The underlying objective of the integration was to make family planning programme more acceptable, efficient and cost effective (Mosley and Sirageldin, 1987). In India, early 1990s decentralization and integration seems to be the similar words underlying the implementation of various social development programmes in India, especially the reproductive and child health programme. It is pointed out that service integration has received renewed emphasis following 1994s International Conference on Population and Development (ICPD) in Cairo (Hardee and Yount, 1995). It was found that the pace of annual progress after 1998 in many RCH indicators is slower than before and a few indicators (child immunization) have worsened, despite the expenditure on the programme being doubled (Srinivasan *et al.*, 2007).

However, the decentralization and integration of programme was often opposed on the ground that integration would increase the work load of health worker, and may not be effective unless mentioned centrally and backed by full time health professional at the delivery level. Large scale experimental studies were carried out in many developing countries to assess the effectiveness of integration of family planning, maternal child health services, other health services (e.g. nutrition, malaria eradication, parasitic central programme) and developmental programmes (female education, income generation). In Harare, the integrated family programme resulted in high acceptance of contraceptives with less inconvenience for the mothers (Verkuge *et al.*, 1990). In Tunisia, the effectiveness of family planning programme was best where operate alone neither much services nor other programmes (Population Council, 1993). It was found that integrated programmes have been less successful than desired (Rosenfield, 1984), especially when they were attempted in settings where the overall governmental infrastructure was weak.

The use of contraceptive reflects the interaction of demand, supply and subjective constraints related to access to contraceptives (Simmons, 1992). Similarly, the use of MCH services reflects the interaction of demand, supply and constraints attached to their realized access. Through a sequence of steps, households (particularly mothers) "perceived the need" (demand) and use the service, a process facilitated by "enabling factors" (supply). Social, economic and structural elements also influence this decision-making process. Cultural and ideational factors affect health behaviour in ways that are similar to their effect on the use of intervention programmes like contraceptive-use and MCH care programmes operate in a similar fashion at the individual and household levels. From a demographic point of view, little information has accumulated on how a woman decides simultaneously to use two types of health care services, one for the survival of her children and the other for controlling her own fertility.

Testing the link between MCH service and contraceptive use with inclusive development programme is difficult. Despite the interest in service integration, methodological problems prevent most studies from explicitly testing whether MCH increases contraceptive use. This problem rests on the many determinants of MCH use that also impact family planning decisions. Many programme managers recognize these joint determinants when they notice that some women are just more receptive to both MCH and family planning services. In statistical terms, endogeneity of MCH services produce biased and misleading estimates of the impact of MCH utilization on family planning receptiveness. Ignoring the independent effect of MCH utilization on contraceptive use requires complex statistical models that allow few studies to test the relationship. Despite the methodological obstacles, the argument in favour of service integration requires evidence that there is a statistical link between MCH utilization and contraceptive use.

Only four recent studies from cross-sectional survey had been investigated to understand this issue and not focusing on the development programme. In multi country study based on Demographic and Health Survey (DHS) data collected from 1986 to 1989, Ahmed and Mosley (2002) used structural equation models to examine the relationship between MCH service use and contraceptive use. They found that the simultaneity between MCH service and Contraceptive use is not explained by the exogenous variables, and suggest that families develop a joint demand for better quality health and limited family size and translate these demands into action by using health services for mothers and for children and by voluntarily regulating fertility. Using 1992 DHS data from Morocco, Hotchkiss *et al.* (1999) investigated that women who more intensely use MCH services are more likely to use family planning services. The study of five countries (Bolivia, Guatemala, Indonesia, Morocco and Tanzania) based on DHS data revealed that in Morocco, Guatemala, and Indonesia, MCH services may be a gateway to improve family planning use, as the intensity of MCH service use was found to be positively associated with subsequent contraceptive use among women, even after controlling for observed and unobserved individual household and community level factors that predispose women to use both types of services (Hotchkiss *et al.*, 2005; Seiber *et al.*, 2005). It is found that none of the cross sectional study have any linkage between inclusive development programme, maternal-child health and contraceptive use. Therefore, this study attempts to examine the effect of inclusive development programme at village level on the relationship between the use of MCH services and contraceptive use.

Data and Methods

The data for this paper have been derived from the District Level Household and Facility Survey (DLHS-3) 2007-08, India (IIPS, 2010). The analysis of this study is based on both individual-level and the service availability at village level data related to the contraceptive use and MCH service utilization. We matched the village level data to each woman by merging them with the individual level data. Women who had at least child under age 5 were included in this study. The World Health Organization (WHO) recommends that immunization be completed by the age nine months. For this study only the last child aged 12-23 months was included because complete immunization history are more likely to be available for such children. In the DLHS, the complete immunization data were available only in the "immunization and child care" section of women's interview schedule was copied from the "immunization cards" and in those cases where card could not be found, immunization status of children was obtained from mothers. Specifically, BCG, DPT (all doses), polio (all doses excluding '0' polio dose) and measles vaccinations were considered in the schedule prepared for the immunization status. Older children are not part of this study because their parents are less likely to have kept the immunization cards; the reverse is likely for younger children. This likelihood may bias the association of immunization with contraceptive use because women with younger children are less likely to use contraceptives. The sample for this objective consisted only of women who had last children aged at least 12 months. However, we assumed that bias would be at a minimum level. The sample for this objective is 9889 women.

Methodological Issues of Structural Equation Models

This research study explains by the economic demand model, which guiding the study assumes that individuals and households make maternal child health (MCH) care and family planning service decisions in order to maximize their total level of welfare. Statistical models for analyzing the relationship between the use of contraceptives and MCH care.

For assessing the structural relationship involving the latent variables, structural equation models (SEM) are generally preferred (Bollen, 1989). We used the general models for estimating simultaneous equation (Joreskog, 1979). In the econometrics literature these simultaneous models are referred to as simultaneous equation models with measurement error or shock error models (Geraci, 1982; Wansbeek and Meijer, 2000). LISREL originally evolved as the full-information maximum likelihoód solution for solving the equation simultaneously. In contrast the two stage probit models are based on limited information maximum likelihood, and solve equation separately. In LISREL framework, we fitted non-recursive models (Berry, 1984; Joreskog, 1979) to take in to consideration the simultaneity between the two latent endogenous variables.

In LISREL standard notation (Joreskog and Sorbom, 1989), the latent endogenous variables are designed by hi1 and latent exogenous variables are designated by x_i for the ith individual. Let COUNTUSE be h_{i1} and MCHUSE be h_{i2}. According to our model, the latent variables are modelled as

$$h_{i1}= b_{12}\, h_{i2} + g_{11}x_{i1}+\dots\dots\dots+ g_{1p}x_{ip}+z_{i1}$$
$$h_{i2}= b_{21}\, h_{i1} + g_{21}x_{i1}+\dots\dots\dots+ g_{1p}x_{ip}+z_{i2}$$

With these two equations, simultaneously exists between the two latent endogenous variables (both h_{i1} and h_{i2} are explained partially by each other) for the ith individual. The random errors, z_i are unexplained components of the latent variable models, have expected values of zero, and assumed to be uncorrelated with the p exogenous variables, z_i. The constant term are absent from the equations because the variables deviate from their means. b_{12} is the coefficient of the structural parameter that indicates the changes in the expected values of h_{i1}, after one unit increase in h_{i2}, keeping x variable constant at their means. Similar the parameter b_{21} indicates the influence of $h_1$on h_2. The b_{12} and b_{21} coefficients are associated with the latent endogenous variables, and x? are associated with the exogenous latent variables.

It is hypothesized that two sets of outcome variables, use of contraceptives and use of MCH services, are interrelated and simultaneously determined. They could not measure them directly but could observe them only through a set of proxy variables. To observe contraceptive use, measured current contraceptive use; to observe MCH care utilization, measured antenatal care, TT use, delivery care by trained health personnel, and immunization of children. Hence, both are latent variables because we could not measure them directly but observed them through a set of related variables. The construction of latent variables is justified for two reasons: (1) instead of dealing with multiple related variables separately, we dealt with a single construct and (2) it was easier to interpret the results because we dealt with fewer variables. It is important to note that the association of a latent variable with another variable reflects the overall relationship of the related observed variables, taken together, to that variable. For example, the association between contraceptives use and one of the MCH care variables may be weak, but taken together, the MCH care variables may have a stronger relationship.

First latent endogenous variable COUNTUSE was constructed directly observed by current contraceptive use without any associated error. Second latent endogenous variable MCHUSE was constructed by antenatal care, institutional delivery, safe delivery, post natal care and immunization for the last children. The latent exogenous variable service availability is usually constructed with (1) distance of health facilities from the centre of village, (2) the availability of community based programmes, mobile clinics, health and family welfare camps in the village, and (3) health facility and provider's availability in village. The analytical framework reproduce the relationship between the latent and observed endogenous and exogenous variables is shown graphically in Figure 24.1.

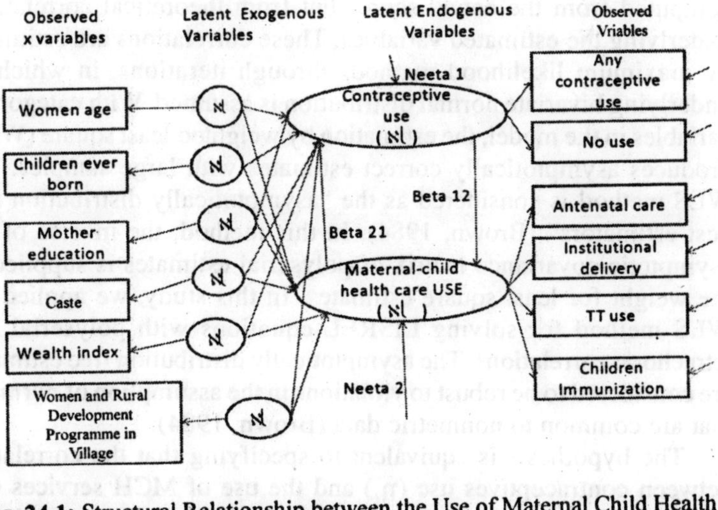

Fig. 24.1: Structural Relationship between the Use of Maternal Child Health and Family Planning Programmes

The construction of latent variables is justified for several reasons: (1) instead of dealing with multiple variables separately, we deal with a single construct; (2) interpretations of the results are easier, because we deal with fewer variables; (3) the result shows the group behaviour, rather than that of a single entity. For example, the association between contraceptive use and one of the MCH care variables may be weak, but together MCH care variables may have rather a stronger relationship. If the analyses are done separately, the relationship could not be discovered in such circumstances.

As noted by Ahmed and Mosley (2002), the estimation of LISREL models by maximum likelihood methods yields inconsistent parameter estimates and inefficient standard errors when the models contain categorical dependent variables, particularly when the estimation is based on Pearson's product moment correlation matrix. To correct this problem and to estimate asymptotically efficient estimates, many methods have been proposed (Bollen, 1989). Because our models also contain categorical dependent variables, we adopted the correction method proposed by Jorsekog (1986). In this method, the correlation among the observed variables are measured by polychoric correlation (when both variables are categorical) and by polyserial correlation (when one of the variables is categorical, and the other is continuous), instead of Pearson's product moment correlation. The polychoric and polyserial correlation are not computed from the actual scores but from theoretical correlations underlying the estimated variables. These correlations are estimated by maximum likelihood methods through iterations, in which an underlying bivariate normal distribution is assumed. With categorical variables in the model, the estimation by weighted least square (WLS) produces asymptotically correct estimates with large samples. The WLS method is considered as the "asymptotically distribution free best estimators" (Brown, 1984). In this method, the inverse of the asymptotic covariance matrix of polyserial estimates is supplied as the weight for least square estimates. In this study, we applied the WLS method for solving LISREL equations with polyserial and tetrachoric correlations. The asymptotically distribution free estimates are considered to be robust to violations in the assumption of normally that are common to nonmetric data (Brown, 1984).

The hypothesis is equivalent to specifying that the correlation between contraceptives use (η_1) and the use of MCH services (η_2) after ξ_i effects are eliminated is not zero. To examine hypothesis,

LISREL model with $\beta = 0$ have been fitted and estimated the covariance between η_1 and η_2 estimating ψ_{21}. The null hypothesis of spurious correlation if the value of ψ_{21} differs significantly from zero is rejected. The correlation between η_1 and η_2 by $\psi_{21}/(\psi_{11}\,\psi_{22})^{1/2}$ has been estimated. The presence of a reciprocal causal relationship between η_1 and η_2 with a non-recursive model specification (Berry, 1984) have been tested. The hypotheses have tested to see whether β_{12} and β_{21} differ significantly from zero and inclusive development programme has an association with health service utilization.

Results of the Structural Equation Model

Uttar Pradesh is the most populous state of India (Census, 2011) and significantly diverges in their socio-economic, demographic, geographic and cultural profiles (NFHS, 2007). Now-a-days Uttar Pradesh has been passing through the third stage of demographic transition, with an estimated death rate of 8.2 per 1000 population and Infant Mortality Rate (IMR) of 63 per 1000 live births (Office of the Registrar General, 2011). A large proportion of the state's population suffers from poverty, with low female literacy and low women autonomy. In the Human Development Index (HDI), it ranked 13th position among the 15 major states of India (Planning Commission, Government of India, 2002) (also see Table 24.1). The distribution of contraceptive use and MCH care use among the women has shown in Table 24.2. It shows there is variability in the utilization levels of services in rural area of Uttar Pradesh.

Table 24.3 shows the characteristics of the women, households, communities and the health services in the rural Uttar Pradesh. The mean age of women is 26 years, 3.4 children per woman ever born and average 7.6 years of schooling are important characteristics of women. At least 25 per cent of women are scheduled castes/scheduled tribes (SC/ST), more than three-fourth women are Hindu and nearly half of women households are poor (first and second quintile) stratum. More than 85 per cent of villages are connected by all weather roads, having health facility and health providers. At least one woman health programme, two adolescent and education related programmes, one rural development programme partially related to health and seven exclusive rural development programmes implemented and have beneficiaries in each sampled village of Uttar Pradesh.

Table 24.1: Socio-Economic and Demographic Profile of the Population of India and Uttar Pradesh

Indicators	*India*	*Uttar Pradesh*
Population (in Millions)	1210	199.5
Density of Population (People/km square)	382	828
Sex Ratio (female/1000 males)	940	908
Decadal Growth rate (%)	17.64	20.09
Crude Birth Rate (Births/1000 midyear population)	22.5	28.7
Crude Death Rate (deaths/1000 midyear population)	7.3	8.2
Life Expectancy at birth, male (in years)	62.6	60.3
Life Expectancy at birth, female (in years)	64.2	59.5
Total Fertility Rate	2.6	3.7
Infant Mortality Rate	50	63
Maternal Mortality Ratio (maternal deaths/100000 live births)	212	359
Female Literacy Rate (%)	64.46	59.26
Per capita Net/State Domestic Product (INR)	46,492	23,132
Human Development Index	—	13

Table 24.2: Percentage Distribution of Contraceptive Use and Maternal Child Health Care Use in the Uttar Pradesh (Rural)

Indicators	*Percentage*
Contraceptive use	
Current contraceptive use (any)	30.8
Current contraceptive use (modern)	16.9
Current contraceptive use (traditional)	13.7
MCH care use	
Antenatal care (ANC)	63.2
Institutional delivery	20.5
Safe delivery	25.7
TT injection received	61.5
Post natal care (PNC) within two weeks	30.8
DPT-3 immunization	37.7
Full immunization	29.3

A complete SEM model is presented in Figure 24.1 of the study. Table 24.4 shows the results to test the null hypothesis that the association between the use of contraceptives and the use of MCH services is spurious/false. The results consistently indicate that the

Table 24.3: Characteristics of Women Households, Communities and Services, Rural Uttar Pradesh

Age (mean in years)	26.2
Children ever born (mean)	3.4
Women's Education (mean in years)	7.6
Religion (Hindu)	84.5
Caste (SC/ST)	24.6
Wealth Index (first and second quintile)	49.8
Village connected by all weather road	87.5
Health facility in Village	87.9
Health provider in Village	94.5
Number of women and children programmes in village (scale 0-2)	1.69
Number of Women (adolescent and education) programmes in village (scale 0-4)	1.83
Number of Rural development (indirectly to health) programmes in village (scale 0-5)	1.28
Number of Rural development programmes in village (scale 0-16)	6.83

Table 24.4: Correlation among Latent Endogenous Variables Contraceptive Use (y_1), and Use of Maternal and Child Health Services (y_2)

Description	Covariance (ψ)	Correlation $\psi_{21} / (\psi_{11}\ \psi_{22})^{1/2}$
Model 1	0.011*	0.054
Model 2	0.015*	0.144
Model 3	0.009*	0.055
Model 4	0.013*	0.154

*$p < 0.05$

use of contraception affects the use of MCH care services (and *vice versa*) independent of common predictor variables. The partial covariance estimate psi (ψ_{21}) between COUNTUSE (η_{12}) and MCHUSE (η_{21}), given the set of latent exogenous variables, were significantly in all selected models ($p < 0.05$). Therefore, we reject the null hypothesis of spurious association ($\psi_{21} = 0$) between the use of contraceptives and the use of MCH care services, which is not due to the common set of predictors. The estimated partial correlation $(\eta_{12}, \eta_{21}) = \{\psi_{21} / (\psi_{11}\ \psi_{22})^{1/2}\}$ was higher in the models 2 and 3. It is suggesting that a higher degree of association between the use of contraceptives and of MCH care services in the rural Uttar Pradesh state. The presence of a significant correlation suggests simultaneity between the uses of two services that is not explained by the exogenous variables included in the model.

Table 24.5: Parameter Estimates of Contraceptive Use (h_1) and Maternal Child Health Care (h_2) from the Structural Equation Models, Rural Uttar Pradesh

Estimates	*Model 1*		*Model 2*		*Model 3*	
	η_1 (COUNTUSE)	η_2 (MCHUSE)	η_1 (COUNTUSE)	η_2 (MCHUSE)	η_1? (COUNTUSE)	η_1 (MCHUSE)
(1)	(2)	(3)	(4)	(5)	(6)	(7)
Lambda Y (λ_y) Estimates						
Current Contraceptive use (Modern)	—	—	—	—	1.000	—
Current contraceptive use (Any)	1.000	—	1.000	—	—	—
Antenatal care	—	1.000	—	1.000	—	1.000
Safe delivery	—	0.176*	—	—	—	—
TT use	—	1.034*	—	—	—	—
Full Immunization	—	—	—	0.781*	—	0.782*
Institutional delivery	—	—	—	1.216*	—	1.216*
Post natal care	—	—	—	-1.242*	—	-1.243*
DPT_3	—	0.302*	—	—	—	—
Beta						
β_{12} (Coefficient MCH)	-0.003	—	0.025†	—	0.014	—
β_{21} (Coefficient COUNT)		0.047*		0.023*		0.031†
if ($\beta_{12} = \beta_{21}$) ≠ 0	0.022*		0.023*		0.028*	

(*Contd.*)

Table 24.5 (*Contd.*)

(1)	(2)	(3)	(4)	(5)	(6)	(7)
Gamma (γ_x) Estimates						
Age	0.007	0.055*	0.018	0.026*	0.006	0.026*
Children ever born	0.059*	0.115*	0.047*	-0.051*	0.040*	-0.051*
Mother education	0.045*	0.047*	0.055*	0.053*	0.040*	0.053*
wealth index	0.038*	0.063*	0.056*	0.067*	0.053*	0.067*
Beneficiary of women development programme in village	0.016†	0.014†	-0.017	0.018*	0.009†	0.014†
Beneficiary of rural development programme	-0.010	0.002	-0.005	0.013*	0.001	0.013*
Model Fitness Indexes						
Goodness of fit index (GFI)	**0.998**		**0.989**		**0.988**	
AGFI	**0.992**		**0.946**		**0.942**	
Root Mean Square	**0.083**		**0.100**		**0.101**	

*$p < 0.05$, † $p \leq 0.10$

Instead of fitting the fixed model, attempts have been made to find the best fit model in different contexts. However, for this reason the models are quite different within state. If inclusion of any exogenous variable with insignificant statistics $p > 0.10$ leads to a poor fit, the variable is excluded from the final models. The results of the best fitted models are presented in Table 24.3. The coefficient β_{12} relate the influence of the use of MCH services on contraceptive use, and the reverse relationship is shown by the coefficient of β_{21}. Only model 2 has a significant reciprocal relationship between the use of contraceptive and of MCH care services, independent of the common determinants. In model 1 ($p < 0.05$) and Model 3 ($p < 0.10$), coefficient of any and modern contraceptives use significantly increased the use of MCH care services. However, the coefficient of MCH care services use was positively associated with the contraceptive use, but the relationship was not statistically significant. The result indicates that the use of any and modern method contraceptives significantly affected the use of MCH care services but lacked of reverse association.

Discussion

We examined the common factors that determined the use of family planning/contraceptive methods and of MCH care services in rural Uttar Pradesh, particularly the simultaneity between these two behavioural outcomes. It is found that there is no specific pattern of common exogenous predictors except children ever born ($p < 0.05$), mothers education ($p < 0.05$), wealth index of households ($p < 0.05$), and beneficiaries of women development programme ($p < 0.05$), for the use of contraceptives and MCH care services. It can be also viewed that use of these two services is simultaneously determined in rural Uttar Pradesh. It is mention that families develop a joint demand for improving the quality of their health and for children. It could translate these demands into action by adopting contraceptive methods and by using health services for children and for mothers simultaneously. However, it could also be possible that women experience with either the health or family planning systems removes the barrier to using other. Model 2, the use of any contraceptive methods and use of MCH care services are simultaneously related and statistically significant, independent of other intervening covariates. In the other two models, contraceptive use significantly influenced the use of MCH care services. Because the β_{12} and β_{21} are close in most setting, we also fitted the models with alternative specifications that $(\beta_{12} = \beta_{21}) \neq 0$. The results of showed that the bs are statistically significant ($p < 0.05$) in all models.

• The study provides empirical evidence for the research that the utilization of one health services influences the acceptors to use other health services, independent of other factors. These findings has not only theoretical value but also important for policy implications. The finding particularly suggests that if both family planning and MCH care services are introduced and available, the likelihood of adoption of both is higher, compared to if only one is available. The rationale for synergies of family planning and MCH care services has been based on three grounds: improving the efficiency and effectiveness of services; meeting clients need from "one-stop" service; and accelerating the pace of health and demographic outcomes. This study further substantiate the rationality of synergy, it is likely that if both kinds of services are available, the use of both will be increased, compared to if only one type of service is available.

The concept of "gateway behaviour"—the one type of behaviour may trigger or facilitate other positive behaviours—less important in the health literature. Used primarily in the United States domestic literature, the approach has been used to investigate substance abuse by considering whether the use of tobacco and marijuana are "gateways" to the use of harder drugs (Ginzler *et al.*, 2003), and health behaviours by considering whether exercise is a potential gateway behaviour to healthy eating (Tucker and Reicks, 2002). As pointed out by Acharya *et al.* (2004), if there are gateway health care behaviours, then an understanding of these relationships can potentially be used by health sector decision makers to improve the efficiency and effectiveness of health services delivery and communication programmes. It has been believed by behavioural scientist that adoption of a new behaviour is a process that involves passage though a sequence of behavioural changes that begin with awareness extends through accumulation of knowledge, and ends in behavioural consolidation. According to these hierarchical behavioral models (McGuire, 1981), the consolidation of one behaviour (e.g. adoption and use of contraceptives) may facilitate more rapid adoption of similar behaviour (e.g. health care use or *vice versa*). The findings of this study further substantiate these behavioural models.

Analysis revealed a good fit of a common model is attempted. Each model has been analyzed that fitted well statistically. Due to the variability of MCH indicators (e.g. antenatal care, institutional delivery, safe delivery, post natal care, TT injection, DPT-3 and full immunization of children), grouped in two type of endogenous variable. First endogenous type included antenatal care, institutional delivery, post natal care and full immunization of children; second endogenous type included antenatal care, safe delivery, TT use, and DPT-3 immunization of children are modelled. These two types of endogenous behaviour can be easily looking at in model which shows the importance of indicators and also it looks over at the multicollinearity between the variables.

The caste of women and beneficiaries of rural development programme doesn't influence the contraceptive use. Another side, age of women are a statistically significant ($p < 0.05$) variable which influence the MCH care services in rural Uttar Pradesh. It is observed that the children ever born, mother's education and household wealth index are common determinants which influence the both services. It can also view that beneficiary of implemented exclusive women development programme in villages have a statistically significant impact on the MCH use than contraceptive use. It supports to the assumption that the availability and beneficiary of women health programme with others in village have an influence on the MCH use (Catherine, 2008). But, these programmes are not much significant impact on the contraceptive use. It shows that programmes are more focus on women health which helps to increase institutional delivery, children immunization, children health and adolescent health.

Another area of concern is the reporting of TT use in DLHS sample. If partly protected by earlier TT use, women may not be immunized during the last pregnancy. The survey did not consider a difference between one or two doses of TT. The DLHS is almost similar to the DHS (in India NFHS) pattern and estimates provide at district levels. Moreover, it was found that the validity of DHS data on TT questioned on the ground that many women may not differentiate between TT use and any other injection during the pregnancy period (Buckens *et al.,* 1995). The DLHS data may suffer from recall bias, particularly the maternal care histories that are based on recall of retrospective events which may have occurred as far as five years ago. Moreover, the validity of response on the type of services provider for delivery care, particularly in distant past, is not evaluated.

Another challenge of this study was to analyze the data with multiple binary outcomes variables with endogeniety. Now-a-days, there is a growing interest in analyzing binary outcomes variables containing endogenous variables (Guilkey *et al.*, 1995). Although some estimation methods are available in different disciplines especially in psychology, the properties and behaviour of these estimators are still under investigation and have limitations in practice. The SEM method which was originally developed for continuous scale variables is used in this study. Although recent implication of these models allows binary/categorical variables, and hold asymptotic properties, there is a lack of Monte Carlo evidence for judging their performance compared to other estimation procedures. As a part of empowered action states (EAG), Uttar Pradesh is under increasing financial constraints in national and state budget for health sector. Therefore, to expand social services requires efficient and effective allocation of scarce public resources. This study attempts to provide information on the determinate of utilization of health and family sector can be more informed, effective and planned.

It is recommended that policymakers can advance support for service integration in the long run, offering service together can save money and better serve the public health needs. It is often useful to work with private providers to expand access to health services. At times, non-governmental organizations can serve women more easily than the public sector can or women may prefer these services. Policymakers can also encourage corporate social responsibility programmes to integrate services. Such programmes should be designed to address time and cost barriers that prevent the most economically disadvantaged women from accessing MCH and contraceptive services. Working across sectors to achieve mutually reinforcing objectives such as universal secondary education for girls and integrated services will build consensus for reform.

In conclusion, family planning is an underutilized, but powerful tool for helping countries as well as states makes progress towards the Millennium Development Goals (MDGs). Providing Maternal-child health care with contraceptive services is a more effective way than separate services to reach women who need family planning. Through higher utilization of contraceptive, more families are lifted out of poverty, fewer mothers and children die unnecessarily and more women reach their full potential for contributing to society and the economy. It is found that there is need to do more focus on modern method particularly spacing method and there is no association of rural development programme with MCH and family planning services. As primary health care programmes expand, there is an increase in demand for services and for quality of care which leads to considerations of structural changes in delivery systems, changes in communication and information strategies, integration and coordination of health services with other developmental programmes. This study advocates for the synergies of family planning, MCH services through inclusive village development programme. It may be benefited to programme planners and policy makers to implement a programme in a more effective manner in Uttar Pradesh.

REFERENCES

Achraya, K., Maxwell, K.A., Middlestadt, S.E. and Storey, J.D. 2004. *Gateway Behaviors and Factors: Improving the Effectiveness and Efficiency of Health Communication Programs.* Washington, DC: USAID Health Communication Partnership.

Ahmed, S. and Mosley, W.H. 2002. "Simultaneity in the Use of Maternal Child Health Care and Contraceptives: Evidences from Developing Countries." *Demography* 39(1): 75-93.

Anderson, R. 1968. *A Behavioral Model of Families Use of Health Services. Research Series No.25.* Centre for Health Administration Studies, University of Chicago.

Berry, W.D. 1984. *Nonrecursive Causal Models. Sage University Paper Series on Quantitative Applications in the Social Sciences* 07-037, Beverly Hills, CA: Sage.

Bollen, K.A. 1989. *Structural Equation with Latent Variables.* New York: John Wiley & Sons.

Browne, M.W. 1984. "Asymptotic Distribution Free Methods in Analysis of Covariance Structures," *British Journal of Mathematical and Statistical Psychology* 37: 62-83.

Buekens, P., Tusi, A., Kotelchuck, M. and Degraft-Johnson. 1995. "Tetanus Immunization and Prenatal Care in Developing Countries," *International Journal of Gynecology and Obstetrics* 48(1): 91-94.

Catherine, T. 2008. *The Right Messages – to the Right People – at the Right Time, World Vision India Flexible Fund Case Study.* Accessed at www.flexfund.org on Aug. 26, 2011.

Faruqee, R. 1982. "Integrating Family Planning with Health Services: Does it help?," World Bank Staff Working Papers 515. World Bank, Washington, DC.

Geraci, V.J. 1982. *Simultaneous Equation Models with Measurement Error.* New York: Garland.

Ginzler, J.A., Cochran, B.N., Domenech-Rodriguez, M., Cause, A.M., and Whitebeck, L.B. 2003. "Sequential Progression of Substance use Among Homeless Youth: An Empirical Investigation of the Gateway Theory," *Substance Use and Misuse* 38: 725-758.

Guilkey, D.K., Bollen, K.A., and Morz, T.A. 1995. "Binary Outcomes and Explanatory Variables: Tests and Solutions with an Application to the Demand for Contraceptive Use in Tunisia," *Demography* 32(1): 111-131.

Hardee, K. and Yount, K. 1995. *From Rhetoric to Reality: Delivering Reproductive Health Promises Through Integrated Services.* North Carolina: Family Health International.

Hotchkiss, D.R., Rous, J.J., Seiber, E.E., and Berruti, A.A. 2005. "Is Maternal and Child Health Service Use a Causal Gateway to Subsequent Contraceptive Use?: A Multi-Country Study," *Population Research and Policy Review* 24: 543-571.

Hotchkiss, D.R., Magnani, R.J., Rous, J.J., Azelmat, M., Morz, T.A. and Heikel, J. 1990. "The Effects of Maternal-Child Health Services Utilization on Subsequent Contraceptive Use in Morocco," *Journal of Biosocial Sciences* 31(2): 145-165.

International Institute for Population Sciences (IIPS). 2010. *District Level Household and Facility Survey (DLHS-3),* 2007-08: India. Mumbai: IIPS.

Joreskog, K.G. 1979. "Structural Equation Models in the Social Sciences: Specification, Estimation and Testing," in *Advances in factor analysis and Structural Equation Models*, edited by K.G. Joreskog and D. Sorbom. Cambridge, MA: Abt Books.

Joreskog, K.G. and D. Sorbom. 1989. *LISREL 7: A Guide to the Programme and Application.* Chicago: SPSS.

McGuire, W. 1981. "The Theoretical Foundation of Campaigns," In Rice W. and Paisley W. (eds.) *Public Communication Campaigns*, Beverly Hills: Sage.

Mosely, W.H. and I. Siregeldin. 1987. "The Effects of Family Planning on Improving Efficiency and Effectiveness of Health Services," Paper Presented at the International Conference on Better Health for Women and Children Through Family Planning, Nairobi, Kenya, October 5-9.

Office of the Registrar General (2007). *Sample Registration System (SRS),* Statistical Report 2007, Office of the Registrar General, Government of India, New Delhi.

Planning Commission, Government of India (2002). *National Human Development Report: 2001*, Yojana Bhawan, Sansad Marg, New Delhi.

Available: http://www.developmentfirst.org/india/planning_comission/HDI_report2001.pdf. Accessed 2011, July10.

Population Council. 1993. "Family Planning for Couples in Rural Areas—Tunisia," In *Operation Research Family Planning Database Project Summaries*. New York: Population Council.

Rahman, M., Mosely, W.H., Khan, A.R., Chowdhury, A.I. and Chakraborty, J. 1980. "Contraceptive Distribution in Bangladesh: Some lessons Learned," *Studies of Family Planning* 11: 191-201.

Renke, W. 1985. "The Lampang (Thailand) Health Development Project: Proposed Methodology for Integrated Cost/Task Analysis," In *Evaluating Family Planning Programmes: International Experience with Cost Effectiveness and Cost Benefit Analysis*, pp. 351-390. edited by I. Siregeldin, D. Salkever and R.W. Osborn. New York: St. Martin's Press.

Seiber, E.E., Hotchkiss, D.R., Rous, J.J., and Berruti, A.A. 2005. "Maternal and Child Health and Family Planning Service Utilization in Guatemala: Implications for Service Integration," *Social Science and Medicine* 61: 279-291.

Simmons, G.B. 1992. "Supply and Demand, Not Supply vs. Demand: Appropriate Theory for the Study of Effects of Family Planning Programmes on Fertility", pp. 59-77 in *Family Planning Programmes and Fertility*, edited by J.F. Phillips and J.A. Ross. Oxford, England: Clarendon Press.

Taylor, C.E., R.S. Sarma, R.L. Parker, W.A. Renike, and R. Faruquee, eds. 1983. *Child and Maternal Health Services in Rural India, The Narangwal experiment: Integrated Family Planning and Health Care,* Vol.2. Baltimore, MD: Johns Hopkins University Press.

Tucker, M. and Reicks, M. 2002. "Exercise as a Gateway Behaviour for Healthful Eating among Older Adults: An Exploratory Study," *Journal of Nutrition Education and Behaviour* 34: S14-S19.

Verkuyl, D.A., S. Fawcus, Willis, and H. Mutusa, 1990. "Integration of FP into Maternal and Child Health and Other Services at Harare Central Hospital," *Central African Journal of Medicine* 36: 311-315.

Wansbeek, T. and E. Meijer. 2000. *Measurement Error and Latent Variables in Econometrics*. Amsterdam, North Holland: Elsevier Press.

Recommendations

International Workshop on "Innovation, Infrastructure & Village Information System for Development Planning" 10-11 Nov. 2011

Distinguished scientists, academicians and social worker, participated in the International workshop unanimously recommended following:

- Natural resources qualitative information for physiographic, drainage system, rainfall, land use, soil, wastelands, water resource, forest and vegetarian cover should be collected at village level based on Khasra Map.
- Innovated natural infrastructure information should be geo-referenced on village Khasra Maps (GPS).
- Village information system should be developed using geo-spatial technologies and are create awareness and training to use the database for local area planning process.
- To innovate natural infrastructure at village level for developing rural India with participation of local planning bodies.
- To create natural infrastructure information system through land sat imaging at 1:5000 scales.
- Social infrastructure information system must be updated every year especially for the population, caste composition workplace and educational profile etc.
- Economic infrastructure information for the agricultural production, land ownership, cultivation primary, secondary and tertiary sector of own should be collected and collected every year at the household level for the rural India.
- Institutional infrastructure developed at village level through government, Semi-government and Non-government agencies from historical and religious to development.

Irrigation, road, immunization, drinking water, energy, medicine, health, agro-industry, small scale industry, animal hospital, banking, education, rural development etc. should be collected updated at village level.

- Ongoing implementation of rural development programmes implemented by government and Non-governmental organizations and fund from peoples representatives etc. should be evaluated and progress report should be monitored.
- Need-base village development planning with the environment of local people should be done and village development planned programmes should be implemented with grass root level village information.
- Problems of the villagers and its solutions should be solved within the village itself and achievement of the action work and constraints of any should be resolved.
- Natural, socio-economic institutional infrastructure use at village level should be innovated and village level information system must be prepared to fulfil the increasing demand of fuel wood, fodder fibre, fruits, fisheries, food grains and vegetables etc. for cattle and human population.
- Innovation of infrastructure and information system for strengthening women empowerment programmes and rural health programmes.
- Innovation of infrastructure and village information system should be develop by the people, for the people and implemented with the people at village level in west rural India.
- To review and examine the substantive scope for innovation-information-infrastructure at grass root level.
- To critically examine the need base village development planning.
- To strengthen Panchyati Raj System and the powers, rights, financial, administrative, managerial support system her self government to in sense sustainable development.
- To identify the challenges, problems, constraints for information, innovation, infrastructure and village development planning.
- Suitable remedies should be suggested for innovation of infrastructure and villager information system for development planning with the continued/sustained participation of local people for local area planning and implementation of need-base programmes at grass root level.

Index

Aam Admi Bima Yojana, 260
Accredited Social Health Activists (ASHAs), 272, 291, 294, 296
Administrative power, 202
Afforestation programmes, 91
African development, 40
African Socialism, 200
African traditional knowledge systems, 47
Agricultural development, 257, 263
Agricultural operation, 66, 145, 148, 152
Agricultural Organization, 37
Agricultural University, 148
Agriculture Department, 165, 184
Agri-Horti-Afforestation programme, 82, 84, 92
Agro-climatic zone, 167
Agro-ecosystem health, 145
Agro-forestry systems, 62
Ainu tribes in Japan and Maori tribes, 59
Akan model, 196
Albuquerque declaration, 57
American Cancer Society (ACS), 278
Amethi Development Block, 187, 248, 250
Anand Agricultural University (AAU), 162–64, 169
Anganwadi Centres, 156, 188, 247, 250
Animal hospital centre, temporary, 188
Animal Service Centre, 246
Antyodaya Yojana, 187
Aquifer zone, 10
Arable land, 30, 36, 68
Assiduous development, 180
Assistant Nursing Midwife (ANM), 272, 291
Asymptotic covariance matrix, 304
Attention-deficit/hyperactivity disorder, 278
Authoritative models, 253

Banjar land development, 85
Banking facility centre, 158
Bank-supported programmes, 44
Bargain power, 129
Better Traffic Management system, 280
Bharat Stage III Emission Norms, 279
Biological Diversity Convention, 254
Biomedical models, 289
Black gram, 30, 63, 65–67
Brahma Kumari Spiritual University, 149
Breeding programmes, 254
Brong Ahafo Research and Extension Centre (BAREC), 199
Brundtland Commission, 43
Burigoalini Union of Satkhira district, 97

Canal irrigation system, 10
Capitalism, 201
Cardiovascular systems, 277
Case detection rate, 295
Caste
- Backward caste, 117, 119–20, 187, 189, 249
- Chamar caste, 116–17
- Chauhan caste, 116–17
- comparison, 213
- composition, 113–14, 117, 120
- Kahar/Kashyap caste, 116–17
- Kori castes, 117
- Yadav backward caste, 117

Central Rural Sanitation Programme, 186
Certification System, 283
Chieftaincy systems, 193
Child development, 157, 184, 215, 221
Child health, 290, 298
Child Safety Scheme, 272
Civil society activists, 44
Civil society organizations, 261
Climate Change Projections for India, 56
Climate modelling scenarios, 59
Climate system, 59
Climate-sensitive sectors, 59
Communication programmes, 70, 271, 311

Community Based Organizations (CBOs), 206
Community development, 205–6, 262
Community health centres, 296
Community Projects Conference, 178
Community Radio Station, 261–63
Compensatory fund for Afforestation (COMPA), 88, 91
Computerization models, 283
Constitutional development, 198
Consumerism, 55
Contraceptives method, 310
Cooperative radio, 263
Corporate social responsibility, 313
Correlation matrix, 81, 304
Cost Credit Limits (CCL), 229
Credit-cum-subsidy programmes, 239
Crop/cropping systems, 167
Cropping systems, 65, 167
Cross-sectional survey, 300
Cultural Organization, 37

Dairy development, 188–89, 215
Data management systems, 290
Day labourer, 22, 29–30, 37
Day wages, 242
Deccan Development Society (DDS), 261
Declaration on Environment and Development, 57
Delivery systems, 314
Democratic development, consolidating, 193
Democratic methods, 233
Demographic survey, 47
Demographic transition theory, 297
Dental education institutions, 270
Department of Rural Development, 235
Designing implementation methods, 60
Development agencies, 80, 89, 197, 200
Development authority, 200
Development departments, 164
Development media theory, 253
Development of Women and Children in Rural Areas (DWCRA), 234–35, 239
Development planning, 180, 233, 317–18
  integrated sustainable, 138
  integrated village, 113
  micro level village, 220
  oriented village, 245
Development policies, sustainable, 196
Development programmes, 3, 44, 113, 157, 183, 186, 236, 297, 300
  exclusive, 297
  inclusive, 299–300, 305
  integrated village, 137
  planned, 114
  rural, 188, 215, 305
  self-sustainable various rural, 183
  social, 298
Development radio, 263
Development workers, 40, 137
Developmental activities, 220
Developmental goals, 220
Developmental planning, 220
Diesel Vehicles' Emissions, 277
Diplomatic missions, 197
Disease Control Programme, 268, 295
District Rural Development Agency (DRDA), 159, 235, 248
Drainage density, 81, 89
Drainage impedance, marked, 7
Drainage pattern, 5
Drainage system, 4–5, 213, 317
Drought Prone Area Programme (DPAP), 234, 239

Earth Summit, 254
Eco-development, 69, 88
Economic demand models, 301
Economic development, 3–4, 11, 54–55, 136, 175, 179–80, 187, 240–41, 253
Economic policy, 234
Ecosystems
  aquatic, 127
  estuarine, 131
  fragile, 63
  restoration, 53, 61
Education and Communication programmes, 271
Education and Multimedia Research Centre (EMMRC), 261
Educational infrastructure, 154, 157
Effective Radiated Power (ERP), 262
E-Krishi Kiran Programme, 162–64, 169
Emission Controls Manufacturers Association, 280–82
Emission Norms for CNG and LPG Vehicles, 279
Emission norms, stringent, 280
Emission standards, 283
Emissions Control, 281
Employment Assurance Scheme (EAS), 234, 237, 240
Empowered Action Group (EAG), 294, 313
Energy development, 186
English Quit India, 152

Environmental determinism, 94
Environmental education, 199
Environmental movements, 256
Environmental Protection Agency (EPA), 207
Eurocentric models of ecological conservation, 44

Family planning programme (FPP), 298–99, 303
Family welfare camps, 303
Family Welfare Councillors, 272
Family Welfare Programme, 246, 269
Farm Yard Manures (FYM), 66, 146, 165
Farming model, 67
Farming-related GHG emissions, 62
Finance Commission, 248, 250
Financial-banking-cooperatives industry, 214
Fisheries, 12, 57, 59, 74–76, 78, 81–82, 84, 88, 92, 96, 105, 107, 122, 126–27, 249–51
Fishermen organizations, 133
Fishing nets, 104, 123, 130
Fishing zone, 130
Five Year Plan, 232–33
Focus Group Discussion (FGD), 13, 15, 99–100, 109, 190–91, 194
Food and Agricultural Organization, 37
Food crops, 17, 65, 67
Food grains, 12, 78, 81, 84, 86, 88, 248–51, 318
Food Scarcity, Coping Pattern in, 22
Forest Department, 100, 103
Forestry plantation, 82, 90–91
Fuel wood, 11–12, 40, 48, 74–78, 81, 84–86, 88, 92, 248–51, 318

Ga Traditional Council (GTC), 201
Gandhi's teachings, 178
Ganga Kalyan Yojana (GKY), 234–35, 239
Gasoline Vehicles' Emissions, 276
Gasoline, 276, 279, 283–84
Genetically modified organisms, 42
Geographic Information System (GIS), 71, 213, 219–20
Geological Survey Report, 238
Geology, 4–5
Ghana Standards Board (GSB), 207
Government subsidy, 240–41
Gram Banjar Bhoomi Vikas Yojana, 88, 90–91
Gram Panchayat, 166, 238–39, 241–42
Grazing land, 10–11, 74, 76
Green gram, 63, 65–67, 146
Greenhouse emissions, 277
Grievance redressal mechanisms, 243
Ground water recharge, 10
Groundwater, 42, 68
Group Insurance Scheme, 188

Health and Family Welfare Programme, 246
Health Care Programmes, 272
Health status, 18–19, 288
Hen/cock farming centre, 186
High Performing States (HPS), 295
Hindustan Aeronautics Ltd. (HAL), 120, 157–58, 188
Holistic analysis approach, 113
Horticultural land in Sarai Khema village, 10
Horticulture growers, 212, 226–27
Horticulture programmes, 188
Human Development Index (HDI), 305
Human development, 192, 199
Human population, 11–12, 75, 78, 81–82, 84, 86, 88, 92, 188, 249, 251, 318
Human resource management, 287, 290–91

Idle emission test, 283
Ignorance, 60, 80, 89, 141, 143–45, 176, 179
Immunization, 272–73, 287, 290, 301, 303, 318
Inclusive governance systems, 206
India Meteorological Department (IMD), 56
Indian Public Health Standards (IPHS), 296
Indian system of medicine, 288
Indigenous Peoples' Global Summit, 58
Indira Awas Yojana (IAY), 159–60, 185, 234, 236
Individualism, 180
Infant mortality rate (IMR), 186, 287–88, 290, 292–94, 305
Information System
- spatial, 142
- special, 137
- villager, 318

Innumerable developmental challenges, 263
Institutional delivery 292, 294, 303, 312

Integrated Nutrient Management (INM), 146
Integrated programmes, 299
Integrated Rural Development Programme (IRDP), 234–35
Integrated sustainable inclusive development, 230
Integration of Knowledge Systems, 60
Intensified JRY (IJRY), 234, 237
Intensive development, 58
Intergovernmental Panel on Climate Change (IPCC), 57, 59
International Conference on Population and Development (ICPD), 298
International Marine Organization, 123
Irrigation system, 91, 137

Janani Suraksha Yojana (JSY), 272–73, 288, 292, 294–95
Jawahar Gram Samridhi Yojana (JGSY), 234, 239
Jawahar Rojagar Yojana (JRY), 234, 236–37, 239
Job Card, 185, 189, 241
Judicial system, 284

Kharanja, 189, 248–49
Kisan Mitra Yojana, 188
Knowledge systems, 40, 60, 70
traditional, 43, 54, 70
Knowledge-intensive sectors, 263
Krushak Kalyan Karyakarma (KKK), 234, 238
Kumasi Ventilated Improved Pit (KVIP), 204
Kutch Mahila Vikas Sangathan (KMVS), 261

Land Management and Land Development, 221
Land ownership, 136–37, 139, 317
Land preparation, 42, 87, 90–91
Land reform programme, 189
Land Reform Schemes, 185
Latrines, 247, 250
Line departments, 86, 189
LISREL models, 304–5
Live births, 288, 292, 294, 305
Local village association, 29
Lord Krishna, 148, 150
Low.cost housing schemes, 159
Low Performing States (LFS), 294–95
Mangrove ecosystem, 128
Mangrove forest, 100, 102
Maori tribes in New Zealand, 59
Marketing support system, 86
Marketing system, 90
Mass Education Department, 237
Mass Emission Norms, 279
Master of Social Work (MSW), 210, 212, 271, 273
Matamatega of Kaumana (MOK), 60
Maternal child health (MCH), 298–301, 303, 314
Maximum likelihood methods, 304
MCH care programmes, 299
Medical education, 270
Medical Health Centres, 188
Medical health programmes, 189
Medical support system, 271
Methodological obstacles, 300
Microbial Organisms, 145, 153
Micro-organisms, 83, 95
Micro-spatial scales, 60
Millennium Development Goals (MDGs), 192, 313
Million Wells Scheme (MWS), 234, 236, 239
Minor irrigation department, 246
Miraculous power, 179
Mitigation measures, 129, 280
MNREGA, 91, 185, 189, 241–42, 249–51, 260
Moisture availability index, 162, 164, 166–67, 169
Moneylenders, 29, 100
Mother care programmes, 272
Mother health programme, 272
Movement of Moon and Crop Yield, 151
Multilateral cooperation, 192

Nation and Rural Development in India, 175
National Commission on Agriculture, 89
National Democratic Congress (NDC), 205
National Forest Policy, 89, 91
National Health Policy, 295
National Rural Employment Guarantee Act, 88
National Rural Employment Programme, 236
National Rural Health Mission (NRHM), 215, 250, 260, 267–69, 272–74, 287–89, 291, 294–96, 298
National Rural Livelihood Mission, 260
National Wastelands Development Board, 74

National Wastelands Development Mission, 89
Needy villagers, 91, 186
New Patriotic Party (NPP), 205
No-go-zones, 48
Non-African socialism, 201
Non-agricultural rural development interests, 258
Non-forest use, 88, 91
Non-governmental organizations, 167, 313, 318
Non-recursive model specification, 305
NPK fertilizer, 141, 143
Nutrition and Health Care Programmes, 272

Oil exploration, 122, 124, 126–27, 132, 134
Oil industry, 121–24, 126–29, 132–34
Operation Black Board (OBB), 234, 237
Organic agriculture, 145, 148
Organism, 63, 94–96
  consumer, 94
  innumerable, 254
  interrelated, 95
Organizational structures, 268–69

Panchayat Van Yojana, 88, 90–91
Panchayati Raj Institutions, 154, 157, 184, 216, 239, 245–46, 269
Panchayati Raj system, 216, 229, 318
Participatory methodology, 190, 193
Participatory model, 253
Partnership fishing, 104, 106
Party paraphernalia, 205
Pastoralist livelihood systems, 56
Pasture, 76, 96
Perishability of fruits, 212, 226
Planned development, 233
Planning Commission, 232, 305
Plantation programmes, 91
Policy formulation process, 71
Policy planning, 77, 113, 137, 211
Political campaigns, 205
Population Council, 299
Possibilism, 94
Pradhan Mantri Gram Sadak Yojana, 185, 189
Presidential system of government, 203
Preventive health care systems, 271
Primary health care programmes, 314
Primary Health Centre, 268, 272
Primary health system, 287, 290
Prisons Council, 197
Private Public Partnership (PPP), 270, 272
Project Implementing Agencies (PIAs), 240
Promotional programmes 259
Public transport systems, 280

Qi-treated seeds, 149

Radio Bundelkhand, 261
Radio Frequency (RF), 262
Rajeev Gandhi Awas Yojana, 260
Rajiv Gandhi Charitable Trust's flagship programme, 240
Rajiv Gandhi Gramin Vidyutikaran Yojana, 159, 186
Rajiv Gandhi Mahila Vikas Pariyojana (RGMVP), 240
Regional Climate Modelling (RCM), 56
Religion-based systems, 196
Religious temples, 155
Religious tourist centre, 155
Remote Sensing Applications, 220
Research organizations, 99
Reverse association, 310
Rice intensification system, 66
Rio Summit, 57
Rogi Kalyan Samiti (RKS), 292
Rural Connectivity Programme (RCP), 234, 238
Rural Development Programme
  progress of, 215
  implementation of, 114, 182–84, 189, 211–12, 318
  information system for, 182
Rural Health Mission, 267
Rural health programmes, 268–71, 273–74, 318
  extension of, 269
  implementation of, 269, 274
Rural health, 267, 269–70, 273, 289
Rural Landless Employment Guarantee Programme (RLEGP), 236
Rural poverty, 236
Rural sanitation schemes, 189
Rural water supply schemes, 189

Safe motherhood programme, 298
Sanjay Gandhi Hospital and Marketing Centre, 188
Saubhagyawati Yojana, 272–73
Scenario analysis, 61

School health programmes, 288
Seismic surveys, 126–27
SEM method, 313
Semi-preservation of fruits, 226
Service delivery systems, 287
SHG movement, 240
Shock error models, 302
Single party systems, 200
Skill development training, 240
Small Scale and Agro-Industries, 249
Small-scale cooperative production of bio fuel crops, 62
Small scale industries, 10, 158–59, 161, 182, 186, 215, 318
Social development, 199, 233, 287, 289
Social forestry programmes, 11–12, 188, 248
Social Welfare Schemes, 188
Socio-cultural models, 289
Socio-economic conditions, 31, 189, 212, 226–27, 230
Socio-economic development, 58, 60, 196, 211
Soil analysis, 162, 165–66, 169
Soil ecosystem, 147
Soil fertility, 63, 65–66, 150, 166
Soil health, 147, 165
Soil Health Card, 163, 165–66, 169
Spatial data analysis, 220
Species, 63–65, 103, 122, 127, 254
Spurious association, 307
State agricultural universities (SAUs), 261
Statistical models, 300–301
Structural Equation Model, 300–301, 305
Sub-health centre, 156
Sundarban Mangrove Forest (SMF), 93, 97–99, 104–8, 110
Supply of Improved Tool-kits to Rural Artisans (SITRA), 234–35, 239
Sustainable development
  contemporary approaches to, 198
  defined, 43, 252
  inclusive, 210
  integrated, 138
  partnership for, 201
  rural, 255, 257
Sustainable Yogic Farming (SYF), 144–45, 148, 153
Swarnajayanti Grama Swarojagar Yojana (SGSY), 234–36, 238–39

Technology for Emissions Control, 281
Time line analysis, 15
Tourism, 59
Trade unions, 261
Traditional councils, 195
Traditionalism, 191, 194
Training of Rural Youth for Self Employment (TRYSEM), 234–35, 239
Training-cum-Production Centre, 226

UN Millennium Development Goals, 192
Uncultivated land, 8–9, 12, 248
UNICEF cooperation, 235
Union Parisad, 28
Universal secondary education, 313
Universalisation of primary education, 237
Uttarakhand, 64, 289, 295

Vehicular emission reduction, 276
Vehicular emissions, 276–77, 283
Vehicular Pollution Control Programme in India, 283
Video instructional programmes, 257
Village Health Nutrition Day, 272
Village Information System (VIS), 162, 169, 210, 212, 215–17, 220–21, 227, 229–30, 267, 317–18
Village Panchayat powers, 250
Violent campaigning, 205
Voluntary organizations, 261

Wage-employment Programmes, 236
Warabandi system, 68
Wastelands development, 76, 221
Wastelands Programmes, greening of, 91
Water drainage, 247, 250
Watershed Associations and Watershed Committees, 240
Watershed development, 237
Weighted least square (WLS), 304
Welfare associations, 197
Women and Child Development, 184, 215, 221
Women development programme, 310
World Commission, 252
World Health Organization, 37, 301

Youth welfare, 184, 221

Zero tillage for wheat cultivation, 147
Zila Panchayat, 183, 242